BBC Micro:bit
Tests Tricks Secrets Code

5x5 LEDs in new uses:
2 dice as one display
5-digit counter to 99999
Voltmeter to 3.3V
Oscilloscope

Examples using some additional components

Serial Communication to PC

Burkhard Kainka

Copyright © 2016 Burkhard Kainka

www.elektronik-labor.de

All rights reserved

Translation and copyright for English text:
Juergen Pintaske, ExMark
STEM Ambassador

ISBN-13: 978-1092154277

FOREWORD

Micro:bit is a small microcontroller learning system, developed by the BBC in collaboration with the University of Lancaster for seventh grade students in Great Britain. The hardware and software tools are very well suited for work in school. Students can program interesting applications around a 32-bit ARM controller with very little effort, and without the need to worry about details of the hardware involved. As you can see on the Micro:bit web pages, they are very detailed and well used.

But the Micro:bit can do more! It is a complete development system and in addition a versatile single-board computer for all kinds of tasks. This controller can also be used as a measuring instrument in the electronics lab. It is therefore exciting to examine the different properties of the system more closely.

The aim of this book is to explore some of the many possibilities of the Micro:bit. The result of our little expedition into hard and software is something like a complete overview on the topics of microcontrollers, programming, electronics and measurement technology. Many of the aspects also apply to other microcontroller systems or to electronics in general.

I hope you enjoy the experimenting and programming, leading to success with your own projects later!

Some additional material and updates can be found at www.elektronik-labor.de .(now, mostly in German)

Burkhard Kainka, December 2016

Version 1.4

CONTENTS

1 Introduction	6
2 Getting Started using the Block Editor	11
3 Event Control	21
4 Analog measurements	30
5 Static display option	37
6 Measurements at the Port pins	50
7 The Micro:bit oscilloscope	63
8 Characteristic curve measurements	80
9 Micro:bit and the Touch Editor	87
10 Micro:bit and MicroPython	94
11 Programming Experience Toolkit (PXT)	99
12 Programming with Mbed	104
13 Mbed and MicroBit.h	112
14 Micro:bit short circuit protection	116

Burkhard Kainka

1 Introduction

The Micro:bit is supported by a variety of different compilers to generate the code that then runs on the processor. The easiest to use is the Microsoft Block Editor. Existing pre-defined blocks are pushed together, and the program is finished.
In addition, there are other languages ranging from Python via JavaScript to C ++, with each step getting more professional and complex.
I would even say that the Micro:bit is interesting not only for education and school, but equally for universities and, of course, for the own Electronics Lab.

This book sets a clear focus on programming with blocks. It is amazing how far you can get. All essential functions and structures are possible, and for most experiments they are sufficient. The other languages will be only be mentioned briefly here. The aim is to show in examples, how far you can get using the Micro:bit.

But this book is not primarily about programming, it is more about the hardware, electronic experiments and measurements. I want to show that the Micro:bit board can replace many measuring instruments in the electronics laboratory: from a voltmeter to the amperemeter, and the signal generator to a simple oscilloscope, everything is possible.

These different measurement options are not carried out for their own sake; they will help to better understand the characteristics of the system and how the external components are used. With the increasing complexity of computer technology, there is a risk that the basic physical fundamentals are not looked at anymore - and as result one might fail with small things such as the correct control of an LED. But both can be combined: electronic basics and programming.

Often, schools are afraid of electronic experiments because they are associated with a high cost for materials. If each lab seat must be equipped with all the tools and measuring devices, it will lead to considerable cost. But the Micro:bit comes with its own measuring instruments. And therefore, only very few external components are needed here for our examples.

To perform all experiments suggested in this book, you only need the following inexpensive components, which you might have already:

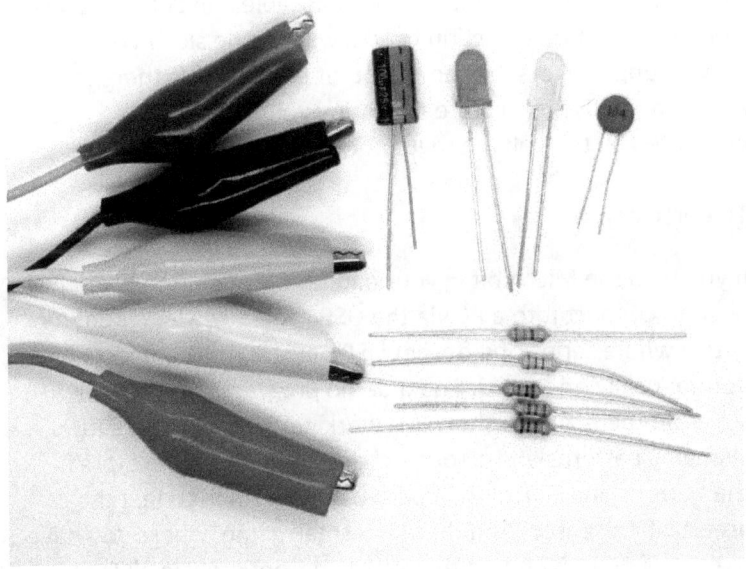

- 5 Crocodile clip cables
- 2 LEDs, one red and one green
- 6 Resistors: 47 Ω, 100 Ω, 330 Ω, 1 kΩ, 3.3 kΩ, 10 kΩ
- 1 Electrolytic capacitor 100 µF
- 1 Capacitor 100 nF

All examples are deliberately restricted to the 5 large connection points, which can be contacted using the crocodile clips. Three ports are sufficient to carry out all essential experiments. Also, a battery compartment is not necessary if the board is only operated via the USB cable and gets the power from the PC. This restriction to USB will on one side avoid all power supply errors, but on the other hand ensure there is always a relatively accurate operating voltage of 3.3 V, generated by the voltage regulator on the board.

The first step

If you hold the Micro:bit in your hand for the first time, you just need to connect it to a PC via the USB cable to switch it on. As with a whole series of different USB devices, they are installed automatically on the PC. And a demo program immediately starts on the board to show it is working. The LED field shows changing patterns and scrolls a character display. Next, press the buttons and swivel the board back and forth to test the acceleration sensor. And there is a small game, where you have to catch a point by tilting the board appropriately in all 4 directions.

Micro:bit – tests, tricks, secrets and code

Microsoft PXT (Beta)

Our new micro:bit Programming Experience Toolkit (PXT) editor provides a programming experience supporting both a block-based editor and JavaScript, along with great new features like peer-to-peer radio.

Try the PXT beta

Python

MicroPython is a completely text-based editor, perfect for those who want to push their coding skills further. A selection of 'snippets' are on hand to help auto-complete trickier tasks and a range of premade images and music are built-in to give you a helping hand with your code.

Try the beta editor with radio

Microsoft Block Editor

The Block Editor is a visual editor and provides an introduction to structured programming via drag and drop coding blocks that snap together. You can also convert a Block Editor script into a Touch Develop script which helps with the transition to text-based programming.

Start with this editor   Documentation   Lessons

Javascript

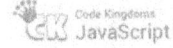

Code Kingdoms is a visual JavaScript editor. It has a drag-and-drop interface making it accessible to beginners. You can also change from the visual editor to a text-based editor which supports the transition to text-based programming as the learner's coding skills progress.

Start with this editor

Microsoft Touch Develop

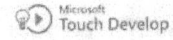

With its touch-based interface, Touch Develop has been designed for mobile devices with touchscreens. It can also be used with a pc, keyboard and mouse. Touch Develop introduces a statically-typed scripting language with syntax-directed editor. It can be used to produce web-based apps that can run online on any platform.

Start with this editor

Mobile Apps

The micro:bit mobile app lets you send code to your micro:bit using Bluetooth wireless technology. No connecting leads needed! Just make sure that your micro:bit is powered up and within easy reach of the phone or tablet running the app. Learn more about using micro:bit on mobile here.

The language selection is located at http://microbit.org/code/

But of course, you want to program something yourself. The introduction is very simple, as during development the greatest importance was placed on the targeted use in schools. Micro:bit behaves like a USB memory and is visible as a drive under Windows. It contains two files: DETAILS.TXT and MICROBIT.HTM. If you open the HTM file using a browser, you are redirected to the Micro:bit web page. This was originally the site www.microbit.co.uk/

In the meantime, the home page has changed to http://microbit.org/, and the starting point for all of the available programming languages is http://microbit.org/code/. Now you can choose a language and get started.

Common to all languages is one fact: they work completely on the web and for this reason need Internet access to http://www.mbed.org. The compiled program is returned as a hex file and then you only have to copy it into the Micro: bit storage medium.
As soon as a new hex file appears in the Micro:bit folder, the controller is automatically programmed with this file. And it means as well, that you can save any generated hex file on your own PC in another folder and transfer it from there to the Micro:bit, without a need for the Internet.

The Micro:bit is usually supplied with power via the same USB port, which is also used for programming. Alternatively, you can also connect an external 3 V battery pack and then even send the software to be programmed wirelessly via Bluetooth. The current consumption is only about 3 mA in idle mode and about 10 mA if all LEDs are switched on, not discharging the battery too quickly.

2 Getting started with the Block Editor

Whenever you start with a new microcontroller, you first write a very simple program. Often it is to flash an LED. And as the LEDs already exist on the board, it can be done in this case without any additional external hardware. This is not quite the same as a simple LED flash program using only one port connection. Here, a whole LED matrix is controlled; in the background the controller has to run a complex program for the multiplex control of these LEDs. But for you it is quite simple to activate. As a first step, we will switch all 25 LEDs on and off at the same time.

The simplest programming environment for Micro:bit is the Microsoft Block Editor. You can get started by just looking at the available blocks and trying them out. In this case, you will use the graphical block "*show leds*"; you can click on the LEDs you want to light up. In this way, you can design a flashing small picture or a logo rather than using all 25 LEDs.

The "clear screen" block is used to switch them all off. In addition, there is the wait command "pause (ms)" and an endless loop "forever". All of these elements can be found in the Basic Group, it includes the elements that you can use for your first, easiest programs. The blocks of the individual groups differ in color, the basic commands are in turquoise.

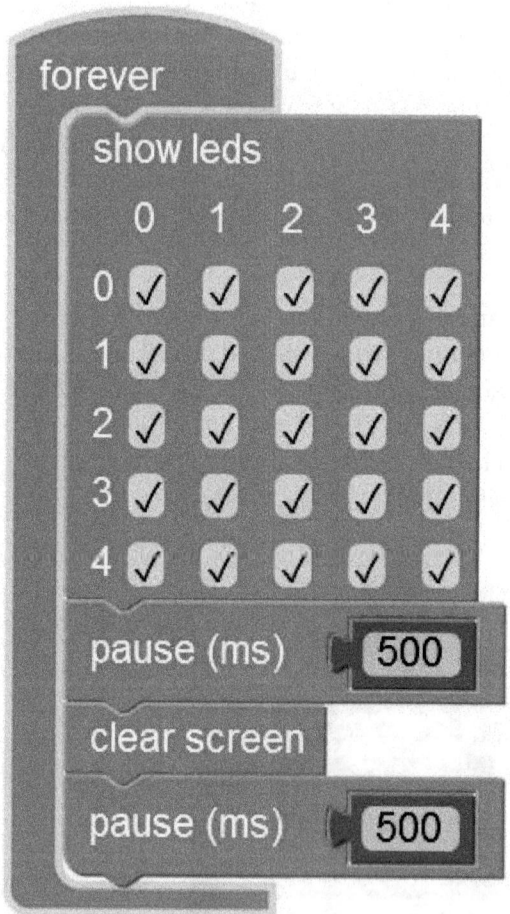

Flash1.jsz

To compile this program, just drag the desired blocks together. Where numbers or texts are inserted, you can change and adapt them accordingly; for example the wait command is set to 500 ms. Then this program is compiled, transferred and starts running.

All programs described in this book can be loaded directly from the author's homepage: http://www.elektronik-labor.de/Microbit/Praktikum.html . The block programs use the suffix jsz. In addition, the compiled hex files of these programs can be loaded from there as well; load them directly into the Micro:bit without any new compilation.

It is very useful to try out all of the available blocks at least once. Using "show number" you display a single digit. If you output multi-digit numbers, they are scrolled through the screen. The same applies to a text output Block "show string".

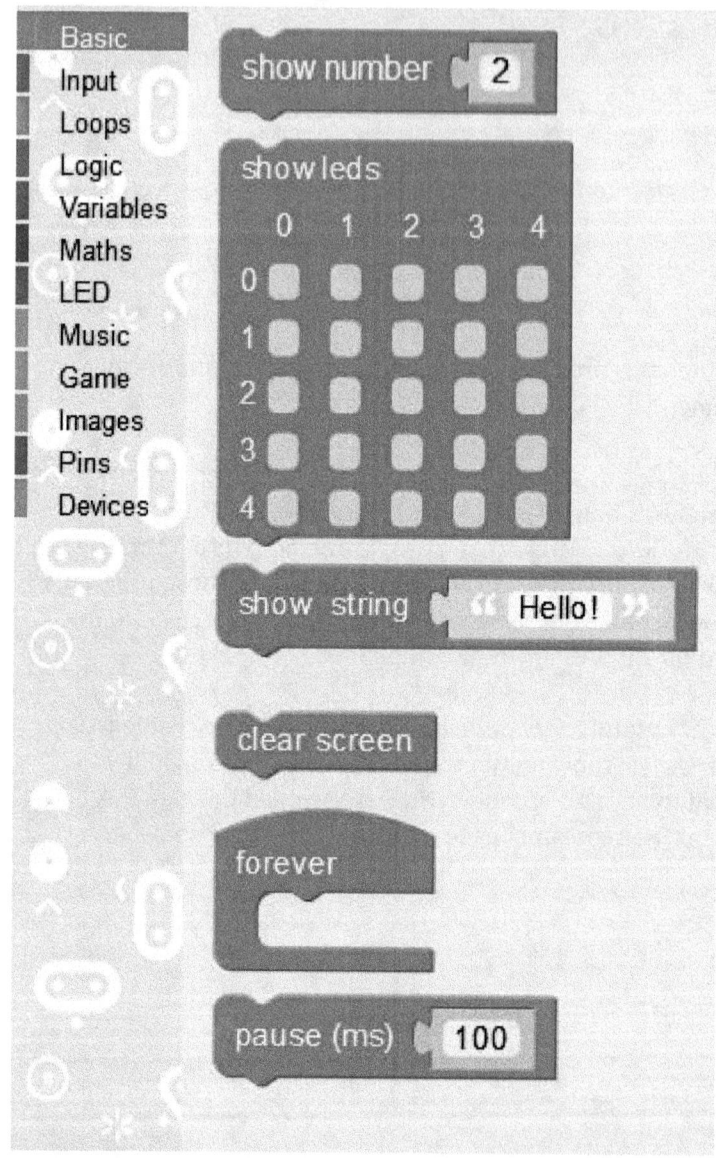

Microsoft Blocks offers a graphical representation for a script, which can also be looked at in text form. Just click on *Convert* and this modifies the view of the program accordingly. Then you end up at Microsoft Touch Develop.

script Flash1 (converted)

function main ()
> basic → forever **do**
>> basic → show leds (▦ , 4 0 0)
>> basic → pause (5 0 0)
>> basic → clear screen
>> basic → pause (5 0 0)
>
> **end**

end function

Flash1-converted.jsz

This script view shows a detail that had been hidden until now: There is a second waiting time of 400 ms for the "*show leds*" command, which is fixed in the block. Knowing this now, it suddenly becomes clear why the flashing was not balanced and why only seven blinking phases are executed within ten seconds. You could change the program and shorten the first waiting time to 100 ms to balance the delays.

With Touch Develop you can also write completely new programs, and some of the options then include expanded possibilities (see chapter 9). In this book, however, the scripting view is mainly used to allow a second glance at the block and possibly allow for easier understanding of a function. Much of the program becomes clearer, and in some cases you then suddenly understand what was meant by the blocks.

LED fading

Many microcontrollers have so called PWM outputs: Pulse Width Modulation of the on and off time; this can be used to change the brightness of an LED with almost infinite resolution. The same you can find in the Micro:bit. But here, in addition, the entire LED matrix can be changed in brightness. This allows for programming of a soft blinker; it means, not just to turn them on and off, but to turn the LEDs brightness up and down slowly. The exact operation of a PWM output is described in chapter 7.

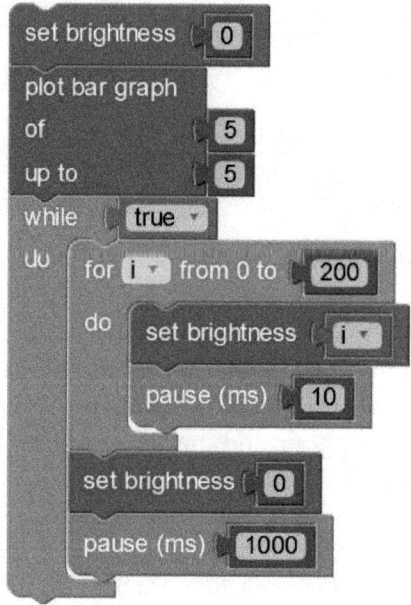

Fading1.jsz

Within the LED command group, there is as well another method for switching all LEDs on. *Plot bar graph* prints a bar with a specified height. In this same group you find "*set brightness*" for setting the LED brightness within the range 0 to 255. In addition, you need a counting loop "for i from 0 to 200" for such a function; the variable name can be defined freely. The counter variable i can be assigned in this counting loop to the LED brightness. The main loop here uses "*while true*", because the *forever* loop has to start at the beginning of a program and does not allow for any previous commands.

```
script Fading1 (converted)
function main ()
    led → set brightness ( 0 )
    led → plot bar graph ( 5 , 5 )
    while true do
        for 0 ≤ i < 200 + 1 do
            led → set brightness ( i )
            basic → pause ( 1 0 )
        end for
        led → set brightness ( 0 )
        basic → pause ( 1 0 0 0 )
        △ micro:bit → pause ( 2 0 )
    end while
end function
```

Fading1-converted.jsz

This code has used a linear change in brightness via incrementing i until now, but as our eye senses brightness in a non-linear way, the brighter phases appear to be a lot longer. Mathematics offers a way out here. Conversions via some the Math's group commands can improve the visual impression.

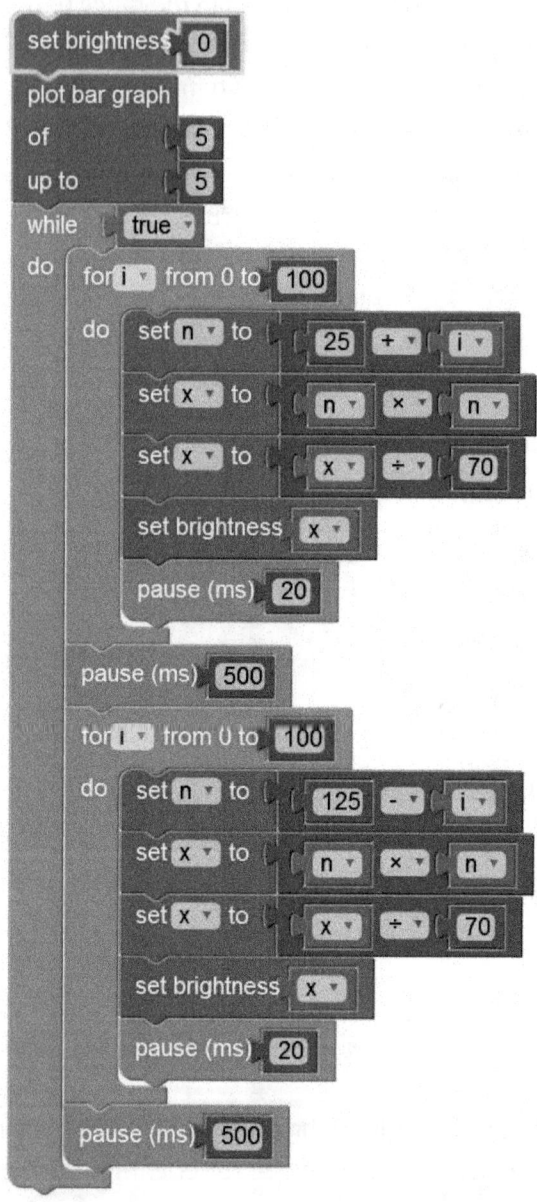

Fading2.jsz (shortened)

This program controls the brightness then following a square equation. The new version now appears as a uniform, well-balanced increase and decrease of brightness.

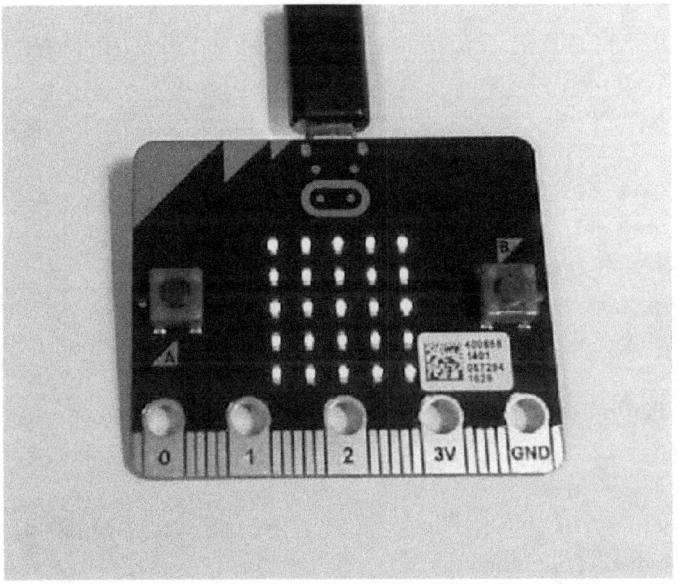

It is interesting to look at the translation of the Blocks into the text form. In this script, all variables are declared at the beginning. And the count loop is executed as a *while* loop.

```
function main ()
    var x := 0
    var n := 0
    var i := 0
    led → set brightness(0)
    led → plot bar graph(5, 5)
    while true do
        var bound := 100
        i := 0
        while i ≤ bound do
            n := 25 + i
            x := n * n
            x := x / 70
            led → set brightness(x)
            basic → pause(20)
            i := i + 1
        end while
        basic → pause(500)
        bound := 100
        i := 0
        while i ≤ bound do
            n := 125 - i
            x := n * n
            x := x / 70
            led → set brightness(x)
            basic → pause(20)
            i := i + 1
        end while
        basic → pause(500)
        micro:bit → pause(20)
    end while
end function
```

Fading2-converted.jsz

3 Event control

A computer should be able to respond to external events. The Micro:bit with its many sensors is particularly well prepared for such activities. The following program shows two very different ways of reacting to external events.

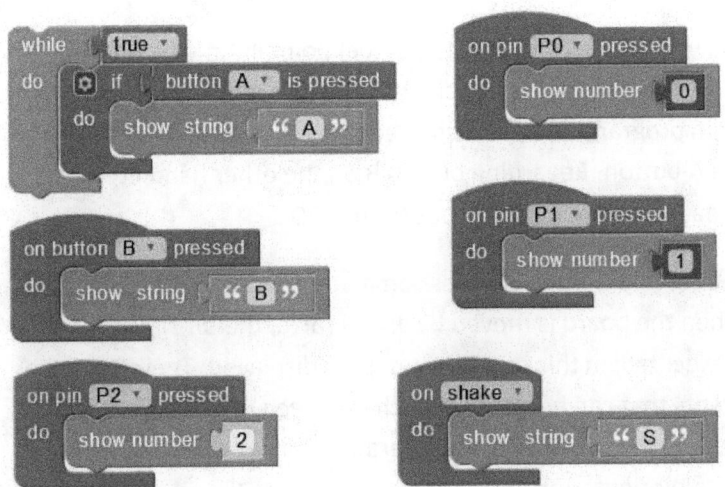

OnPressed.jsz

Our main program runs as an endless loop, and it queries again and again whether the switch A has been pushed and is now closed. If yes, then a corresponding message is displayed on the LED display. This works well, but such an implementation has a crucial disadvantage: the controller is fully occupied with scanning the key even if nothing happens.

Another method is to define event-driven functions which are completely isolated from the main program. They are self-activating independent of the main program loop when such an event occurs. This is followed by an interrupt request to the main program leading automatically to executing this function. The main program is just interrupted briefly to do something else, then returns to its normal function. Pressing button B invokes the *on-button B-pressed* function, then displaying a B.

It is interesting to look at the actual point in time when the execution takes place. Looking at the direct query for A in the main program, the A is displayed immediately when pressing the A button. Regarding button B on the other hand, it must be released before a B is output to the LEDs.

The same program includes some additional input responses. When the board is moved back and forth, the *on-shake* function is executed; in this case, a capital S is displayed. There are other events that can be defined, such *as screen* up or *logo up*, which are also triggered by the acceleration sensor when there is a certain orientation in space.

An event such as on *P1 pressed* is triggered when the connector P1 is connected to the connector GND externally. A switch or push button could be connected here. It is actually sufficient to touch the two connectors directly with two fingers. This works here, because 10 MΩ resistors are connected on the board to the positive supply VCC. Skin resistance is usually in the range of 1 MΩ or lower; as result, the touch changes the P1 input level to a low input voltage.

```
function main ()
    input → on pin pressed(P0) do
        basic → show number(0, 150)
    end
    input → on pin pressed(P1) do
        basic → show number(1, 150)
    end
    input → on button pressed(B) do
        basic → show string("B", 150)
    end
    input → on pin pressed(P2) do
        basic → show number(2, 150)
    end
    input → on shake do
        basic → show string("S", 150)
    end
    while true do
        if input → button is pressed(A) then
            basic → show string("A", 150)
        else add code here end if
        micro:bit → pause(20)
    end while
end function
```

OnPressed-converted.jsz

A dice program

The dice program shows another example of responding to an event. As you can see, there are actually two dice displays. One is started with the A key and the other one via the B key. For both of them, the last number then disappears from the display and the new result appears after one second. However, as discussed before and in this case for key A, the process does not start until it is released, while key B starts immediate reaction.

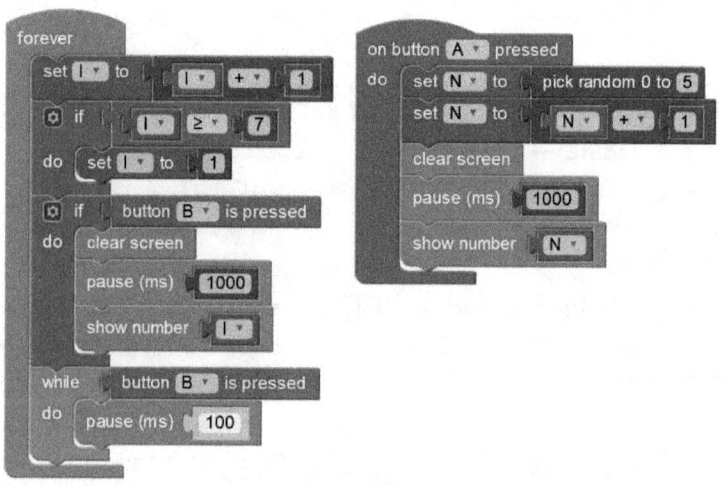

Dice.jsz

Key B is evaluated in the main program. Here, the variable I is continuously increased but limited to a value in the range 1 to 6. Pressing the button takes the current value starts the display output. But then, the program must be stopped in a *while loop* until the push button is released; otherwise you would get more output results in an ascending order.

```
script Dice (converted) 2
function main ()
  var N := 0
  var I := 0
  basic → forever do
    I := I + 1
    if I ≥ 7 then
      I := 1
    else add code here end if
    if input → button is pressed(B) then
      basic → clear screen
      basic → pause(1000)
      basic → show number(I, 150)
    else add code here end if
    while input → button is pressed(B) do
      basic → pause(100)
      △ micro:bit → pause(20)
    end while
  end
  input → on button pressed(A) do
    N := math → random(6)
    N := N + 1
    basic → clear screen
    basic → pause(1000)
    basic → show number(N, 150)
  end
end function
```

Dice-converted.jsz

In this case, the random number is generated by increasing I very quickly, and the actual output of the I value takes place at a random point in time.

The second dice responds only to the A key and is generated via the *on-button A-pressed* function. And in this implementation the random number is generated using the special *random* function Block and then pushed into the possible range of values between 1 to 6. The event function here prevents as programmed internally, that a continuous pressure of the button repeats displays, as the "rolling of the dice" does not start until the button has been released again.

A touch sensor

Using the event *on P0 pressed*, a touch sensor can be programmed as already shown in the first example. However, unfortunately you must touch GND and connector 0 at the same time. A better solution is to connect two cables and provide a larger contact area for each of them.

Our goal was, however, to program a touch sensor which responds even if only one input is touched - without the need for the additional GND connection. The new solution uses an analog input.

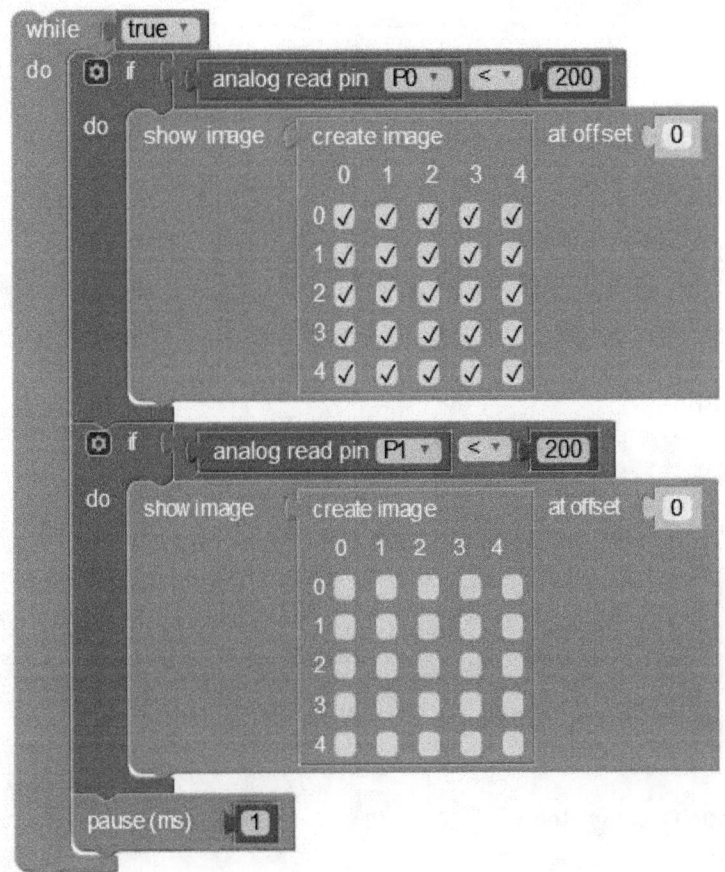

Touch1.jsz

In this case, the input voltage is measured and the program responds to changes of the input voltages when touching either the contact P0 or P1. This then switches between the two pre-defined images. In our example, all LEDs are switched on for the first phase via P0, and all LEDs are off for the second phase using P1. However, it is easy to modify the display and design other images or symbols.

```
script Touch1 (converted)
function main ()
    while true do
        if pins → analog read pin (P0) < 200 then
            image → create image ( ▦ ) → show image (0)
        else add code here end if
        if pins → analog read pin (P1) < 200 then
            image → create image ( ) → show image (0)
        else add code here end if
        basic → pause (1)
        △ micro:bit → pause (20)
    end while
end function
```

Touch1-converted.jsz

Why and how this touch sensor works, is not easy to understand. For a detailed explanation, a few measurements are necessary first; this will be done later (see chapter 7). But at this point already and ahead of time: one can play with the comparison value set to 200. If you make it bigger, then the sensor becomes more sensitive. But from a certain value on, nothing happens anymore, because from then on, the test practically always appears as true and never gets back to the false state.

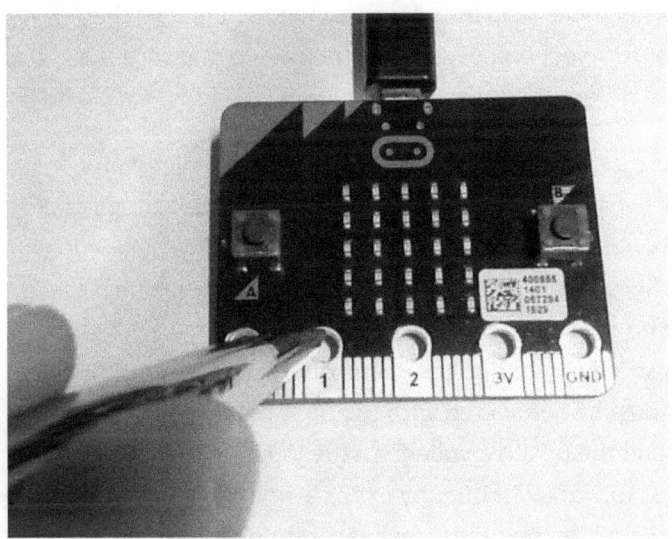

In fact, it is quite difficult to tap the contacts P0 and P1 with a finger without touching any of the adjacent connections at the same time. For practical use therefore, a croco cable should be used to connect to an external contact surface, or use a conductive pen or any other conductive object.

4 Analog measurements

The measurement of electrical voltages requires a voltmeter. And such a function is already built into the controller of the Micro:bit in the form of an AD converter. An analog-to-digital converter measures the voltage and converts it into a numerical value. Often there is an internal switch at the input of the AD converter to select one of several channels to measure the voltage of different connections.

All of our three connections P0, P1 and P2 are designed to measure separate input voltages. In principle, the voltage of a battery could be measured. But beware the risk! When connecting external voltages, mistakes can happen which can destroy the controller. The voltage to be measured at the three inputs is strictly limited to the range of 0 V to 3.3 V and the voltage must be connected with the right polarity. Ground to Ground and the positive voltage to the input. In the beginning, it is better to work without any external voltage source, and use the 3V connection point. If, however, with external voltages within the allowed range, connect a 10 kΩ resistor in series with the measuring cable for protection.

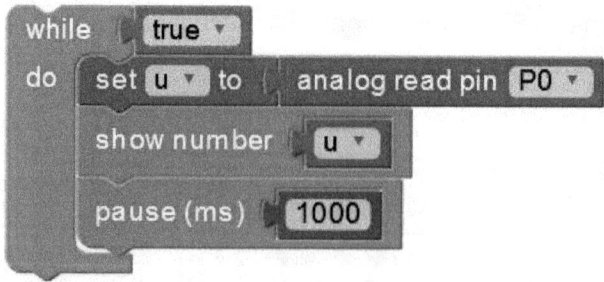

Meter1.jsz

The digital value u of the voltage measured is displayed as a walking pattern via the LEDs. A first test at an open input results in a display of about 238 or 239, with possible slight value variations. Actually, we would have expected a display of zero as no voltage is connected. A "real" voltmeter with inputs not connected would show in such a case 0 V. Could it be, that this open input, as we think, is internally already connected to a voltage level? If a digital multimeter is accessible, we can simply connect it between P0 and GND. The result then shows, for example, 1.7 V. Why this is the case, has still to be investigated.

Next, we connect input P0 via a croco cable directly to the 3 V connector. The LED display now shows the value 1023. This is the largest possible value of a 10-bit number. So, we know now that it is a 10-bit AD converter. All voltages measured are converted into numbers in the range 0 to 1023, with the value 1023 being the voltage at the 3V terminal.

The actual voltage between the 3V hole and GND is not exactly 3.0V, but more accurately about 3.3V if the Micro:bit board is supplied with power via the USB port. The voltage regulator in the processor delivers nominally 3.3 V; there is an additional Schottky diode in series to the 3V connection point. This causes an additional small voltage drop, which depends on the external current drawn.

The numerical value of 238 measured first at the open input thus stands for a voltage of 3.3 V/1023 * 238 = 0.768 V, which is also surprising, because the multimeter had measured 1.7 V.

Let us check now, if at least the voltage 0 V correct? Well almost, as a numerical value of 1 is displayed, which corresponds to a measurement error of approximately 3 Millivolt. But the secret of the open input voltage remains unsolved for now. It will be examined in more detail later (see chapter 12). In advance, however, the following information: On the Micro:bit board, there are resistors of 10 MΩ between these inputs and the 3.3V operating voltage.

This fact, that an open-circuit voltage with a measured value of 238 is visible, is already an important information; this partly explains the behavior of the touch sensor from the last section: if significantly less than 238, a touch can be detected. Why, however, a touch changes the voltage must be examined more closely.

Measurement of temperature

Temperature is measured inside the controller of the Micro:bit and then output directly in degrees Celsius. The accuracy achieved is not very high. Quite a few degrees too much or too little can be displayed. But what is more important: changes of temperature are reliably detected. If the finger touches the processor on the board, a rise of a few degrees can be seen.

```
while true
do  set t to  temperature (°C)
    show number  t
    pause (ms)  1000
```

Meter2.jsz

Measurement of brightness

The LED display serves in this case as a simple brightness sensor. Each LED functions at the same time also as a small photodiode; and so, it is possible to use it as light sensor as a second function. The brightness is measured as a range up to 255. A quick experiment using a laser pointer shows, that the LED in the center of the field is used as light sensor.

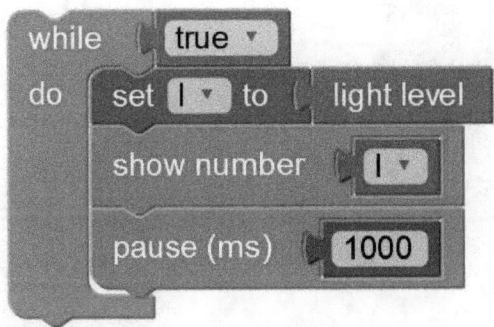

Meter3.jsz

The bargraph display

A simple bar graph display can be used as an alternative, to display the measured value output to the "walking numbers". Although there are only five steps available, they can be programmed as a display with very short response time as is needed for acceleration measurement.

```
while ( true
  do   set a to ( acceleration (mg) z
       set a to ( a × -1
       plot bar graph
       of          a
       up to       1500
       pause (ms)  100
```

Meter4.jsz

The *plot bar graph* function generates the correct scaling automatically if the measuring range is specified correctly. In our case, the measured values given are about -1000 in the resting state, and the sign is initially inversed. The entire display range then reaches 1500, which corresponds to 1.5 g.

Magnetic field measurement

The local magnetic field strength can also be displayed in a similar form. As with acceleration, the desired axis x, y or z must be specified.

```
while true
do  set b to ( magnetic force (microT) z
    plot bar graph
      of
          b
      up to
          100
    pause (ms) 100
```

Meter5.jsz

The first time this program is started, a calibration is automatically performed. You have to move the circuit board around in the air to draw a circle on the display. Subsequently, the smallest changes are displayed which are measured by tilting the board relative to the earth's magnetic field. Moving a magnet closer to the Micro:bit can be measured from quite a distance.

Our simple bar graph display has the advantage of being very fast, but unfortunately only has a very low resolution of five steps high. The problem is as well, that each bar occupies the full width of five pixels. In the following chapter an alternative way is shown; it allows for a large number display option using the same 5 x 5 LED display.

5 Static display option

Scrolling text can also display multi-digit numbers, but there is a problem. You must look at the display for a long time and very concentrated to read a complete text or value. In real life and with the full string visible at the same time you can recognize multi-digit numbers or measuring instruments with a pointer hand at a glance. Could something similar not be possible with the Micro:bit? There are already options to use the LED display to show binary numbers. But this also cannot be read and "translated" easily and needs a lot of concentration. Here, a new idea is used to implement a five-digit decimal display.

The display format

Our display should show five digits in five columns, a result you can read them as usual from left to right. Each digit is represented by a bar of five points, but unusually they are sometimes virtual and mostly outside of the display field.

For a number zero, all LEDs are off as usual. The digits one to five are displayed using one to five LEDs from the lower edge. But then things change. Starting from the number six, our 5 virtual LEDs move upwards and outside the picture, so for the number nine only one LED at the upper edge is switched on. For the number 0, all of the virtual LEDs are below the display, move into it and then out again at the top.

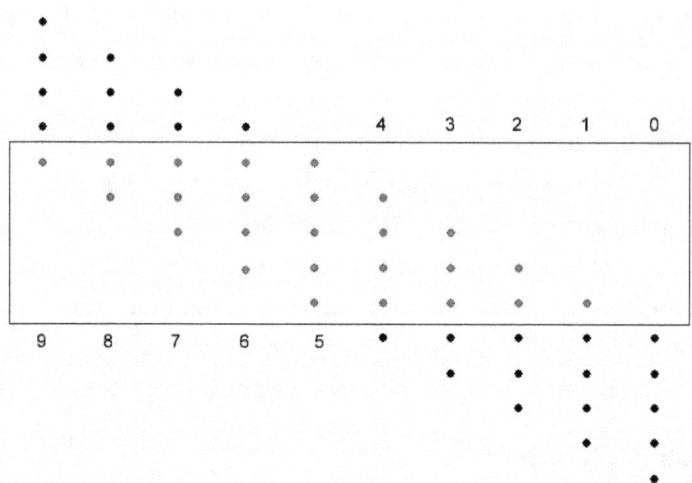

Real and "virtual" LED positions

This representation of the numbers corresponds exactly to the Morse code for digits. Each digit is sent as a combination of dots and dashes, using the LEDs on the Micro:bit; indicator LEDs are turned on for dots and are turned off for dashes. The number 7 is represented in the Morse code with - - ..., the five with ....., and the zero with -----. This comparison came out by accident, but the result proved to be very useful for the LED display.

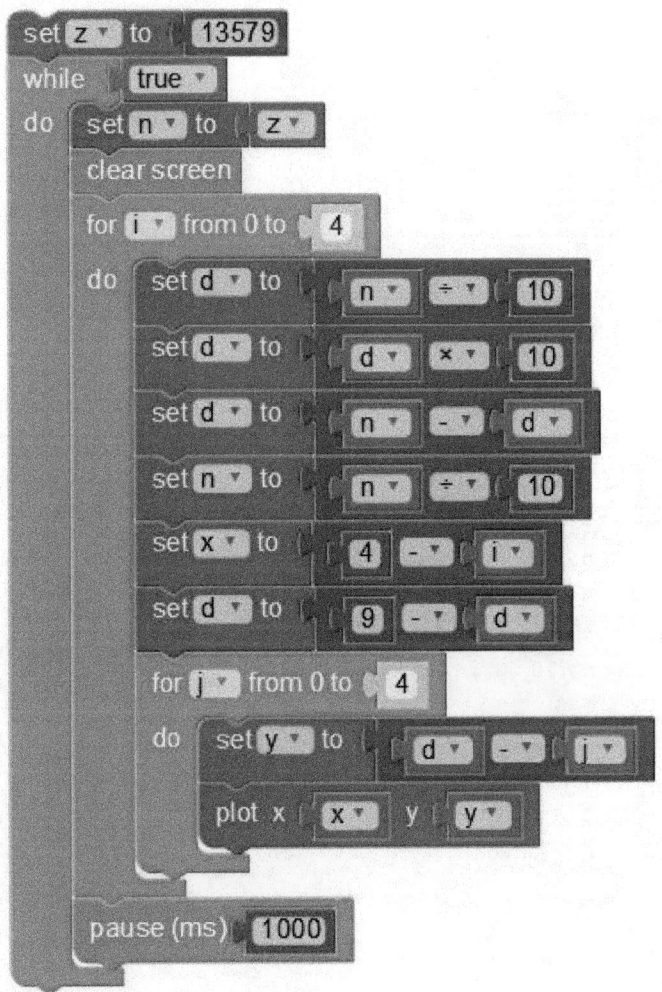

Number13579.jsz

In a first application, a five-digit number will be displayed using the 25 LEDs. The number z = 13579 to be displayed is then copied into n and converted into the required LED pattern.

```
script Number13579 (converted)
function main ()
    var y := 0
    var x := 0
    var d := 0
    var n := 0
    var z := 0
    var i := 0
    z := 13579
    while true do
        n := z
        basic → clear screen
        var bound := 4
        i := 0
        while i ≤ bound do
            d := n / 10
            d := d * 10
            d := n - d
            n := n / 10
            x := 4 - i
            d := 9 - d
            for 0 ≤ j < 4 + 1 do
                y := d - j
                led → plot(x, y)
                    ● y should be between 0 and 4.
            end for
            i := i + 1
        end while
        basic → pause(1000)
        ⚃ micro:bit → pause(20)
    end while
end function
```

Number13579-converted.jsz

The individual LEDs are switched on using *plot x, y*. The LED at the top left has the coordinates 0,0, and at the bottom right the point is 4.4. For each column, exactly five points are switched on, but they can be partially outside the display field.

Our program uses an outer loop and an inner loop. The digits are separated in the outer loop (i = 0 ... 4). It starts with the rightmost digit. As only integers are used here, a division by 10 and a later multiplication by 10 is sufficient to round the last digit down to zero. A subtraction then supplies the last digit as a separate number.

For the next run through the loop, the output number n is then divided by 10. The x position is calculated from the run variable i, and the y position is calculated from the calculated digit d. The inner loop (j = 0 ... 4) creates the bar of five dots. For the 9 on the right, only the LED in the highest row position Y = 0 is visible at the top right. The following four points are above the display field and extend to positions Y=-1, Y=-2, Y-3, Y=-4.

```
d = z           : 13579
i = 0
d = d / 10      : 1357
d = d * 10      : 13570
d = n - d       : 9
n = n / 10      : 1357
x = 4 - i       : 4
d = 9 - d       : 0

j = 0
Y = d - j       : 0
...
i = 4
y = d - j       : -4
```

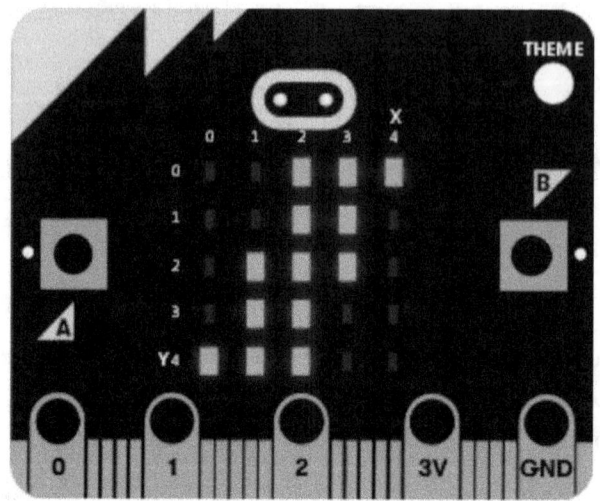

Display of the number 13579

A counter up to 99999

In this application, the variable z is incremented. The result is a running display, where the number increases by ten per second. The highest number is reached after a total of 1000 seconds, i.e. after about 17 minutes. The counter can be stopped at any time by pressing the A button. This makes the program a suitable example for illustrating the new number display format.

Micro:bit – tests, tricks, secrets and code

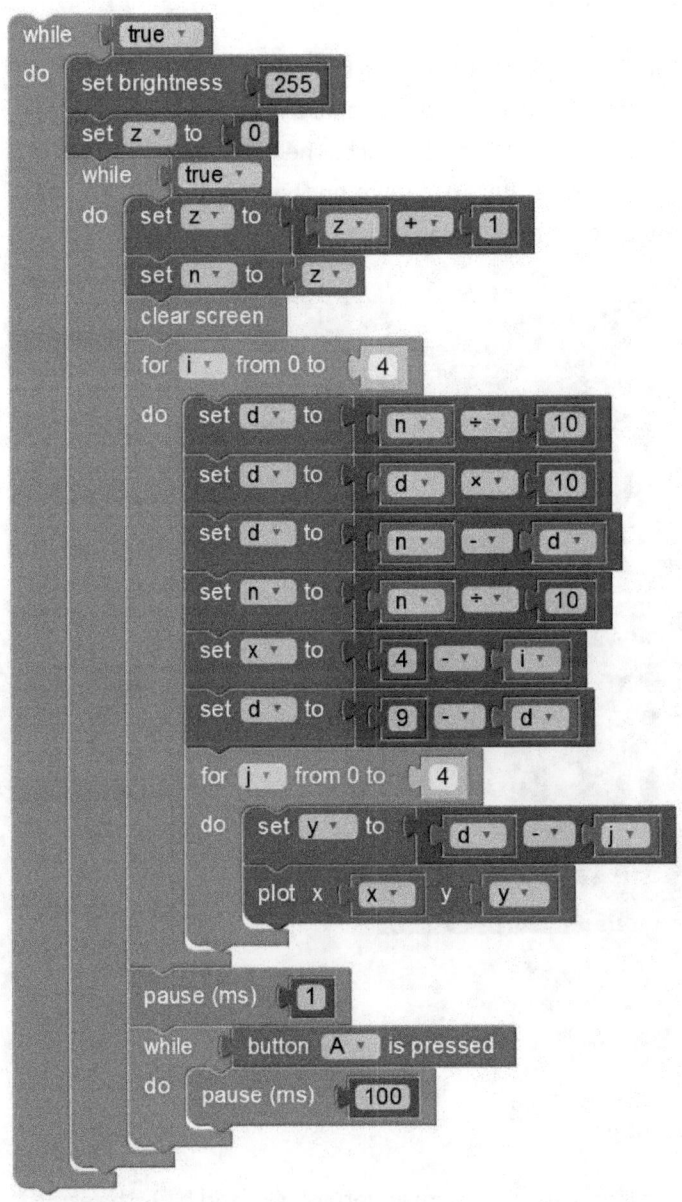

Count99999.jsz

Double event counter

For some tasks one would rather have two small counters than a large number display. This example shows how to generate a double counter with up to 99 each. The pressing of A and B is counted. The A results are shown on the left, B results on the right. And a dark middle column.

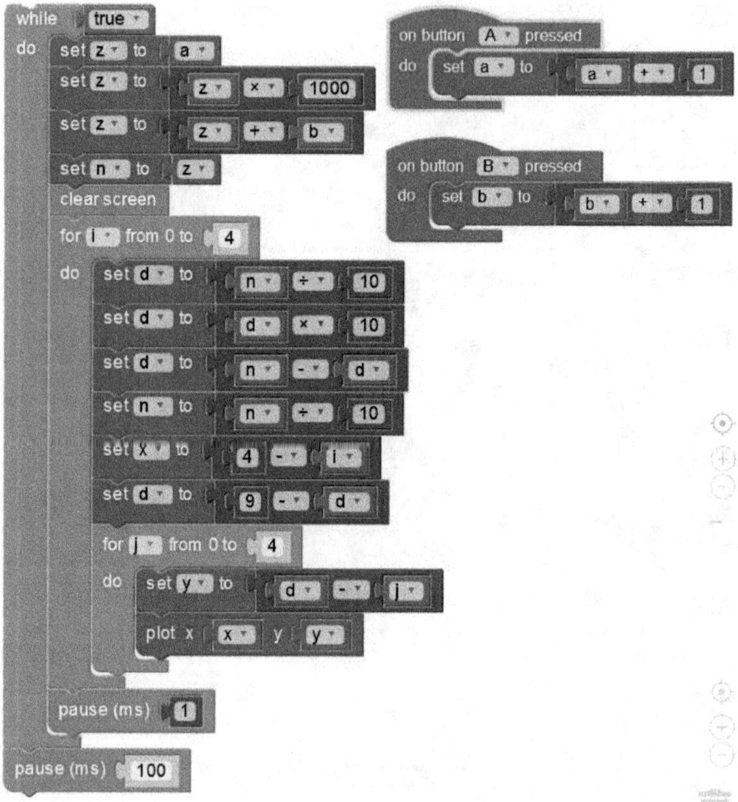

Count_AB.jsz

This program uses two counter variables, a and b. Both results are merged into one number z for display, where value a is multiplied by 1000 to move the result three positions to the left.

script count_Zwei (converted)
function main ()
```
var y := 0
var x := 0
var d := 0
var n := 0
var z := 0
var b := 0
var a := 0
var i := 0
input → on button pressed(A) do
    a := a + 1
end
input → on button pressed(B) do
    b := b + 1
end
while true do
    z := a
    z := z * 1000
    z := z + b
    n := z
```

Count_AB_converted.jsz (shortened)

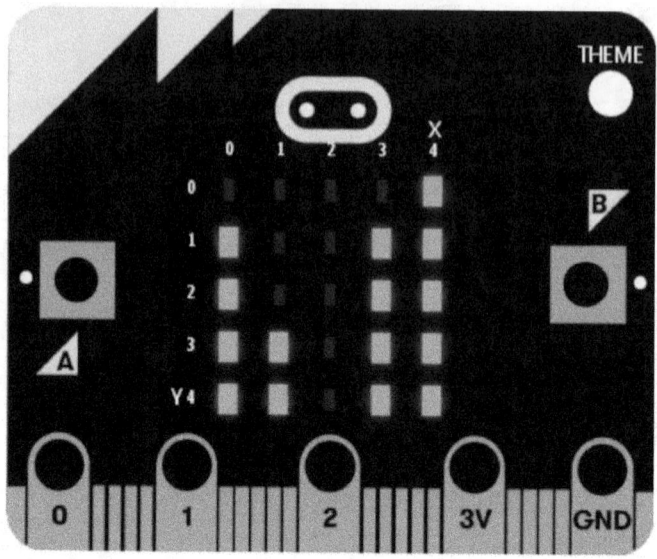

Counter status 42 and 45

Digital voltmeter up to 3300 mV

The Micro:bit AD converter provides numerical values between 0 and 1023, which are voltages between 0V and 3.3V. This raw data can be easily converted into voltages. A multiplication by 3300 and a subsequent division by 1023 calculates the voltage in millivolts, which is then converted to our new display pattern and sent to the LED field.

The analog voltage is measured using channel P0. In addition, the program activates port P1 to active and P2 to in-active, and as result voltages of 3300 mV and 0 mV can be expected. In addition, our new meter can now be used to examine the properties of ports in more detail.

Micro:bit – tests, tricks, secrets and code

```
while  true
do   clear screen
     set brightness  255
     set z to  0
     while  true
     do   set z to  analog read pin P0
          set z to  z  ×  3300
          set z to  z  ÷  1023
          set n to  z
          clear screen
          for i from 0 to  4
          do   set d to  n  +  10
               set d to  d  ×  10
               set d to  n  -  d
               set n to  n  ÷  10
               set x to  4  -  i
               set d to  9  -  d
               for j from 0 to  4
               do   set y to  d  -  j
                    plot x  x  y  y
          pause (ms)  500
          while  button A is pressed
          do   pause (ms)  100
```

Millivolts.jsz

```
script Millivolts (converted)
function main ()
    var y := 0
    var x := 0
    var d := 0
    var n := 0
    var z := 0
    var i := 0
    while true do
        basic → clear screen
        led → set brightness(255)
        z :- 0
        while true do
            z := pins → analog read pin(P0)
            z := z * 3300
            z := z / 1023
            n := z
            basic → clear screen
```

Millivolts-converted.jsz (shortened)

Micro:bit – tests, tricks, secrets and code

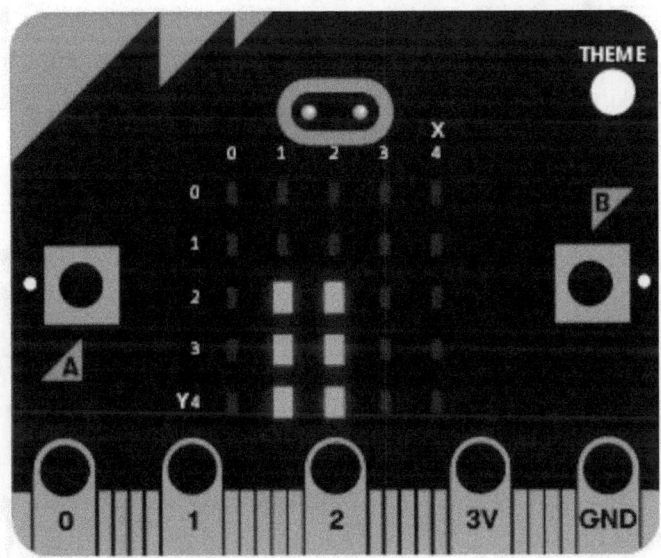

Display of 3300 mV

6 Measurements on the port pins

The three connections 0, 1 and 2 can be used as analog inputs as well as digital inputs and, if required, as digital outputs. You can connect an LED on the outside, usually with a resistor in series for current limitation. But such an output can also switch larger loads like incandescent lamps or motors if an external transistor is used to increase the drive current or voltage.

To evaluate the external connection possibilities of these pins, it is worth taking a closer look at the properties of these ports. How much is a port able to switch, how much current can flow, and from when do you have to worry about something going wrong? These are important questions, especially for experimental work.

A look at the data sheet of the Micro:bit controller nRF51822 shows, that in the setting "high drive" a port can provide up to 5 mA. While an output voltage of 0 V or 3.3 V can be expected without an output load, an internal voltage drop of 0.3 V can occur at 5 mA. Instead of 3.3 V, the voltage in the switched-on state is then drops to 3.0 V. This means still enough, because an LED usually can work using a much lower voltage.

But what happens when there is an overload? Is it possible to connect an LED directly, without a current limiting resistor in series with the LED? To answer these questions, some measurements are useful. To correctly interpret the results of our measurements, we must keep the design of a CMOS output port in mind. It consists of two field-effect transistors, one N-type and one P-type. When the N-FET is switched on, the port is in the low state. When the P-FET is switched on, a high output state is achieved. Each of these transistors behaves like a switch, but additionally has a finite resistance. It is referred to as the on-resistance, because it is valid when switched on.

On the one hand, this resistance leads to a reduction in current and, on the other hand, to a heating up of the transistor.

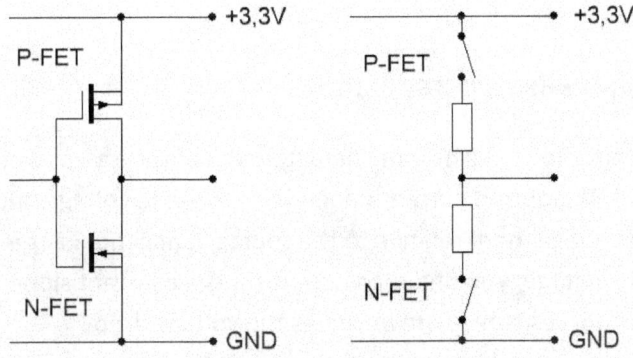

Basic circuit diagram of a CMOS-Port pin

Based on the values in the data sheet we can estimate that the on-resistance is about 60 Ω. However, this resistance is not the same for every load, it increases with higher output current. The output characteristic of a typical MOSFET gives a good impression of this behavior.

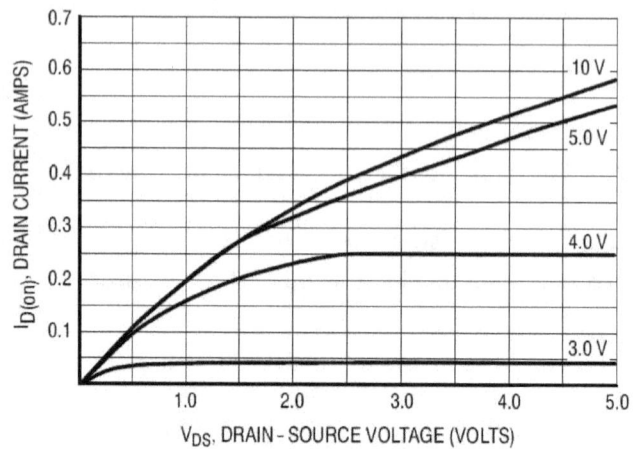

Characteristics of a MOSFET BS107

The diagram shows the output characteristics of the low-power FET BS107. The output current depends on the gate voltage and the drain voltage, both relative to the source connection. For a given gate voltage, you find for small currents a constant slope, i.e. a constant resistance. However, as the voltage drop increases, the curve becomes ever flatter, so the transistor goes into the constant-current mode. Each FET behaves similarly, but the very small FETs in the output pins of a microcontroller are overall higher in impedance and can deliver less current.

The question of possible destruction by overloading is not so easy to answer. For the BS107, for example, a maximum continuous current of 250 mA is specified and a maximum power loss of 350 mW. In the upper region of the characteristic field, however, a current of 0.6 A and a voltage of 5 V are found, i.e. a power consumption of 3 W. For a measurement, the current was therefore only switched on for a very short time to stop the transistor to become too hot via the internal power consumption..

Inside a microcontroller, this looks somewhat different. The loaded FET is part of a larger structure, which must not become too hot. If a single output transistor is loaded more strongly but there is hardly any other power consumption, then heat can be well dissipated, and nothing can happen. Some microcontrollers have a defined maximum overall load on the output ports. For the ATmega328 in the Arduino Uno the maximum current of all ports should not exceed 200 mA.

Looking into the data sheet of the nRF51822 you search in vain for such an indication; presumably, as this controller was never intended for such switching tasks. Typical application areas for this chip are wireless keyboards and mice with battery operation, where the target is always a minimum current consumption. But what a controller can do and what it is supposed to do, is not necessarily the same. One more reason to explore the limits with our own measurements.

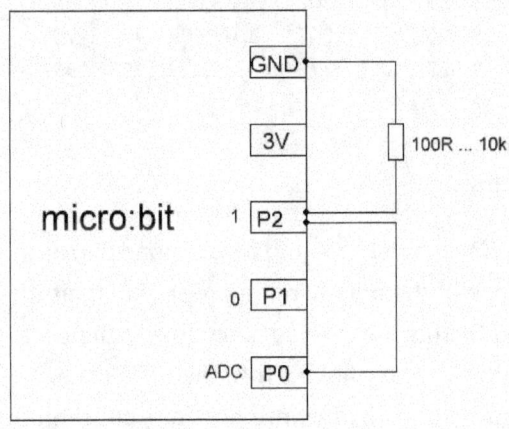

Load of Port pin 2 to GND

For these measurements, different resistors in the range of 100 Ω to 10 kΩ were connected between P2 and GND. The voltage was measured and displayed via an additional connection to the P0 measuring input. From this measured data, the voltage drop in the port, the output current, and the internal resistance can be calculated.

This is where physics and computer science meet. These measurements offer a small course in practical measuring technology. But here no additional measuring instruments are required, as the Micro:bit itself is a measuring instrument. We use the same program as in the last chapter.

	A	B	C	D	E
1	R in Ohm	U in V	3,3V - U	I in mA	Ri in Ohm
2					
3			3,300	0,000	0
4	10000	3,277	0,023	0,3277	70,186146
5	3300	3,235	0,065	0,980303	66,306028
6	1000	3,098	0,202	3,098	65,203357
7	330	2,696	0,604	8,169697	73,931751
8	100	1,522	1,778	15,22	116,81997
9					
10					

Measurement results

In the case of a high load resistance and a small load current, the internal resistance of the port is in the order of magnitude of approximately 60 Ω. This is expected according to the data sheet. However, with currents above 5 mA, the resistance increases significantly. Up to 15 mA were possible, but then the output voltage had already dropped to about half. You can calculate the power converted into heat in the port on the chip from these values. This is up to P = 15 mA * 1.8 V = 24 mW. With a further increase in load, the port suddenly switches off.

This is a very surprising result and showed a totally unusual behavior for a microcontroller. Further tests have shown, however, that this is a built-in security measure, which is only contained in the Blocks, but not in the other available programming languages. Later in chapter 14, this behavior is examined in even more detail.

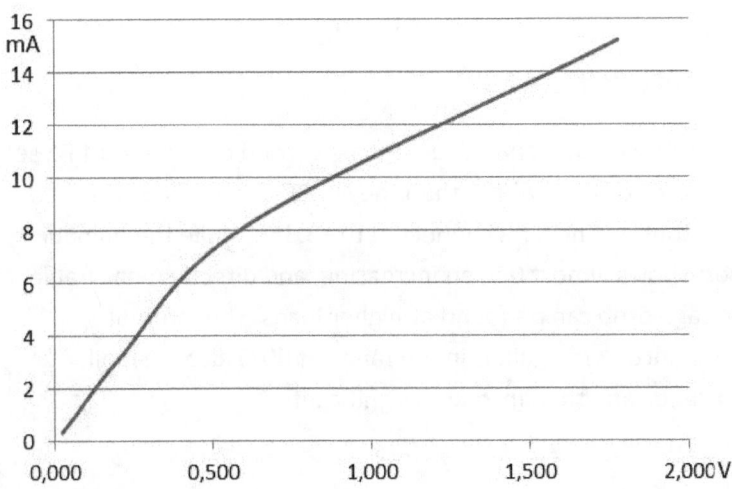

Characteristics of the P-FET

If you plot the current against the voltage drop in a diagram, you can already see the similarity with the typical FET characteristics; but compared to a BS107 with much smaller currents.

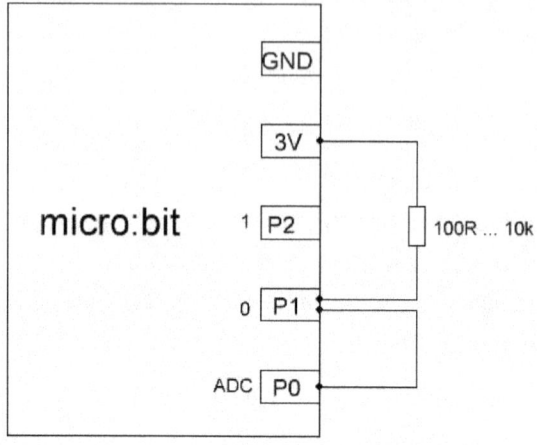

Load of a Port to + 3.3V

The same measurement can also be carried out with port P1 set to the zero state, where then the N-FET is examined. The load resistor must now be connected to +3.3 V. While the unloaded port shows almost 0 V, an increasing, and directly measurable voltage drop can be found at higher loads. The internal resistance here again is in the range of 60 Ω also at small currents, and then increases significantly.

	A	B	C	D	E
1	R in Ohm	U in V		I in mA	Ri in Ohm
2					
3		0,000		0	
4	10000	0,019		0,3281	57,909174
5	3300	0,052		0,9842424	52,832512
6	1000	0,171		3,129	54,650048
7	330	0,561		8,3	67,590361
8	100	1,893		14,07	134,54158

Measurements to VCC

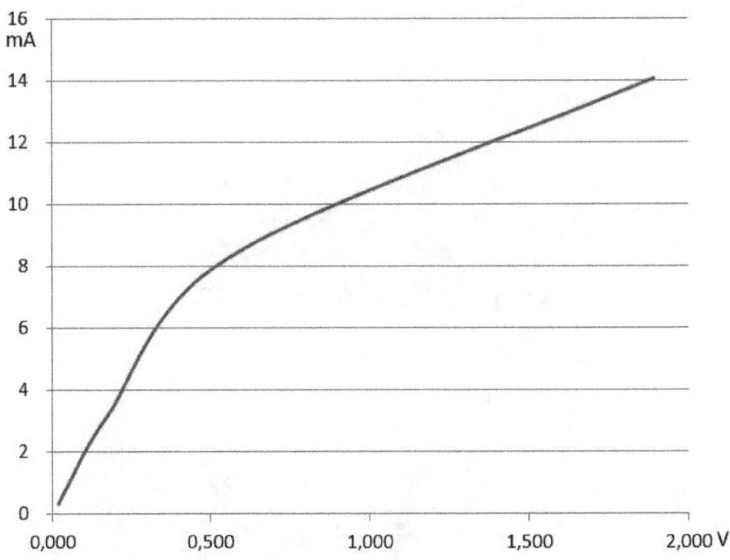

Characteristics of the N-FET

The characteristics of the P-channel FET and the N-channel FET in the output port are very similar. This is desirable, because the port behaves then symmetrically and can drive loads just as well to GND as to VCC.

Measurement of a temperature increase

How much does the controller heat up when the ports are overloaded? For this test, a resistance of 47 Ω was connected between P1 and P2. Thus, we almost have the largest possible load, whereby the power consumption is generated via two ports. A temperature sensing Block was temporarily added to our measuring program. The voltage is still converted but not displayed.

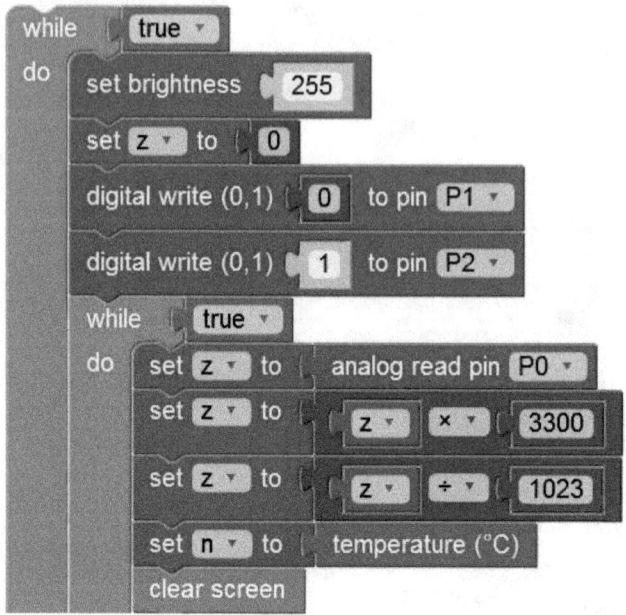

Millivolts.jsz with added temperature measurement

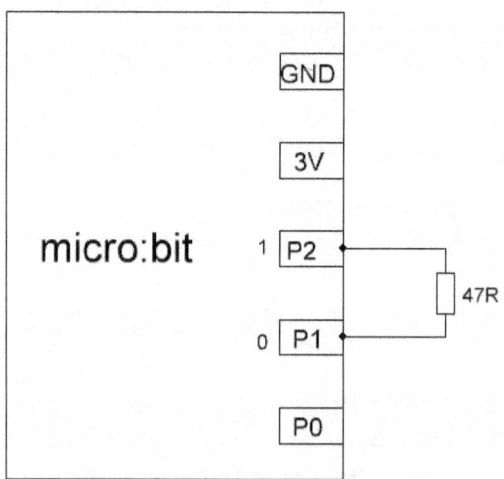

A high load for 2 Ports

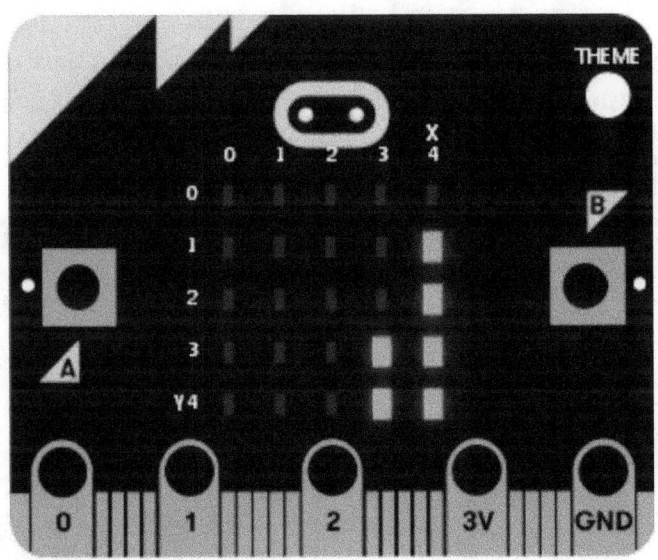

A temperature of 24 ° C

At the beginning of our experiment, a temperature of 24 °C was shown. And the ambient temperature measured as 21 °C. The controller temperature was therefore raised via the normal load by 3 degrees. Now connecting the load resistor, the temperature began to rise, reaching a temperature of 27 °C after a few minutes, three more degrees. This shows that no damage should be expected. A silicon chip is only damaged at a crystal temperature of about 150 °C. Although there may be local overheating in special cases, a risk is excluded in our case if the chip temperature is still well below 100 °C.

From these measurements, we can judge that it is easily possible to connect an LED directly to the port without current limiting resistor. The LED voltage depends strongly on the color used. But practically all modern LEDs operate at voltages in the range of 1.8 V to 2.5 V. The internal resistance of the port must then still absorb 0.8 V to 1.5 V. The LED current is then usually in the order of 10 mA, which is still within the safe range.

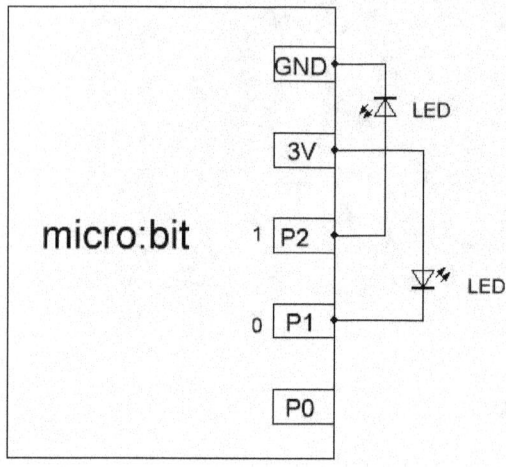

LEDs connected without current limiting resistor

An LED is usually connected via the cathode to GND and with the anode to the port pin. It then lights up in the 1-state of the port. But this also works the other way, with the anode at the 3V connector and the cathode to the port. In this case, the LED lights up in the 0 state of the port. For a simple test, the program Redgreen.jsz was written. The ports are switched on and off using the A and B keys.

```
on button A pressed
do  digital write (0,1)  1  to pin P1
    digital write (0,1)  0  to pin P2

on button B pressed
do  digital write (0,1)  1  to pin P2
    digital write (0,1)  0  to pin P1
```

Redgreen1.jsz

When you connect LEDs between two ports, the doubled on-resistance comes into play and the resulting current is slightly reduced. If it comes to power sensitive applications with battery operation however, then a suitable series resistor should still be used.

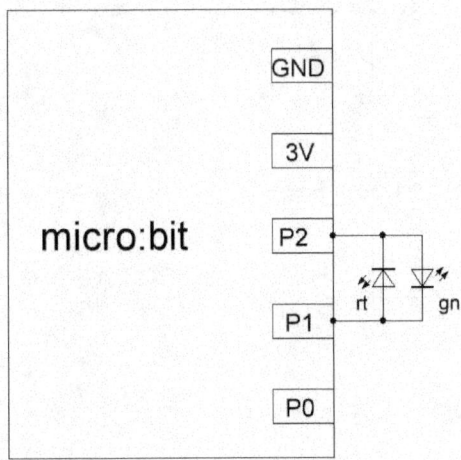

LEDs between two Port pins

Can we control two LEDs with just one port? This can also be done, as the following diagram shows. At first sight one would probably say: Two diodes in the same direction of the current flow directly at the operating voltage, this will fail! But in fact, this is safe, because the operating voltage is only 3.3 V. Both LEDs together would require a little more than 3.3V, until there is a current flows at all. More detailed measurements are carried out in chapter 8. In this case, P1 alone decides which of the two LEDs is turned on. The experiment works well with a simple switching program. This circuit can also be operated via a PWM output to change slowly from red to green. The required software is also available in chapter 8.

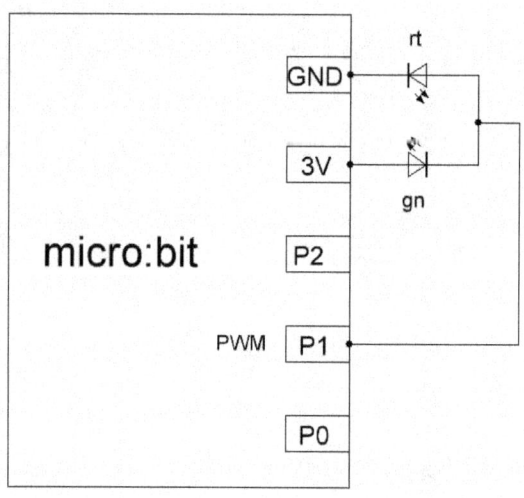

Two LEDs on one Port pin

7 The Micro:bit oscilloscope

The oscilloscope is one of the most important measurement instruments in the electronics lab. It shows the timing of a voltage on a screen. Not only can you see if a voltage is large or small, but also whether, how and how quickly it changes value. A very simple oscilloscope is better than none. And even with only a 5 x 5 LED screen size, measurements are useful.

```
forever
    analog write 512 to pin P1
    clear screen
    for x from 0 to 4
    do  set y to analog read pin P0
        set y to y ÷ 205
        set y to 4 - y
        plot x x y y
        pause (ms) 100
```

Oscilloscope.jsz

This simple oscilloscope performs five measurements and then plots the results from left to right on to the screen. As the AD converter provides measured values up to 1023, the division by 205 reduces the results to the required range of 0 to 4.

In addition, the measured values must be inversed because the zero is on this screen at the top, not the bottom. The individual measurements are plotted directly in real-time, i.e. they are not stored temporarily.

```
script Oscilloscope2 (converted)
function main ()
    var y := 0
    var x := 0
    basic → forever do
        pins → analog write pin(P1, 512)
        basic → clear screen
        var bound := 4
        x := 0
        while x ≤ bound do
            y := pins → analog read pin(P0)
            y := y / 205
            y := 4 - y
            led → plot(x, y)
            basic → pause(100)
            x := x + 1
        end while
    end
end function
```

Oscilloscope-converted.jsz

If Port 0 is connected to a cable, a stray signal can be measured which has the mains frequency. The reason for this are power cables in the area, and voltages in the air are capacitively coupled to one's own body. Without noticing, such an alternating voltage can reach a value of 10 V or more. When the input is then touched, this voltage is slightly reduced, as the connection has a finite internal resistance of approx. 10 MΩ. However, this is usually still sufficient to display a full range on the screen.

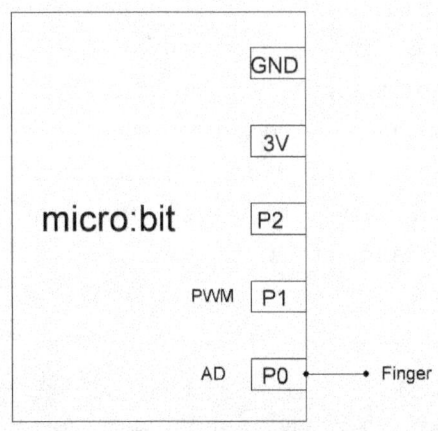

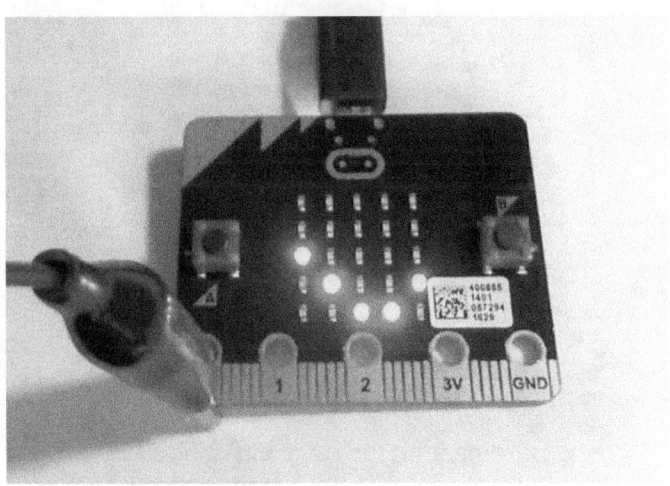

Usually you can see only part of a complete sine wave. But as different measurements show other parts, you get a good impression of the course of a curve.

The fact, that only one part of one sine wave is displayed here is, incidentally, an illusion. A complete oscillation of a 50 Hz signal takes 20 ms. However, our measurement is much slower, because our program inserts a delay time of 100 ms between two measurements. Therefore, dots are recorded from later waves; but if a periodic signal is applied, you can still get a complete picture. This type of a so called subsampling is typical for any digital oscilloscope and can lead to a misinterpretation of measurements. You must always keep an eye on the signal shape and how fast the measurements actually are.

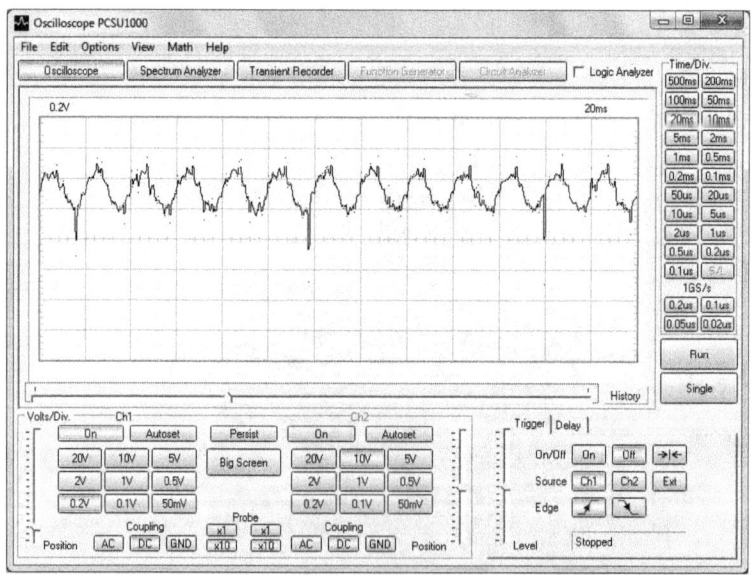

A comparison measurement on the same pin using a fast oscilloscope shows the actual sampling points as small voltage dips in the signal. At these points, the internal AD converter in the Micro:bit takes a voltage sample; it loads the sample-and-hold capacitor which is connected to front of the AD converter. Included in the oscilloscope picture are other noise signals, which get into the measuring cable which is lying open at one end on the table. You can see the 50 Hz signal and as well the progressive scanning points of the Micro:bit relative to the measurement signal.

In this case, our program takes about 102 ms, until the next measuring point is reached. This then scans every fifth 50 Hz oscillation of the 20 ms sine wave. Assuming just for now a sampling period of exactly 100 ms, we would always measure the same phase/position of the signal. This difference of 2 ms will lead to a total of 10 measurements to see a complete wave of 20 ms. The screen width of five conversion points is about half of the 10 points needed, which can be seen in the picture.

It is also possible to draw valuable conclusions from the measurement of such an interference signal when an open input is touched. They can be helpful in other applications. Specifically here, it is about how to write a program that responds to a simple touch of an input. Such a program has already been discussed in chap. 3. There, a measured voltage at the input should be smaller than a limit to detect a touch.

Our measurements using the oscilloscope function, now make it clear what happens. The program simply needs to wait until a negative half-wave of the 50 Hz signal appears and provides a sufficiently low voltage. But this only works reliably if the board has a reference potential via the USB port connected to the PC. When the system is operated via a battery without a set potential, the interference voltages are significantly lower.

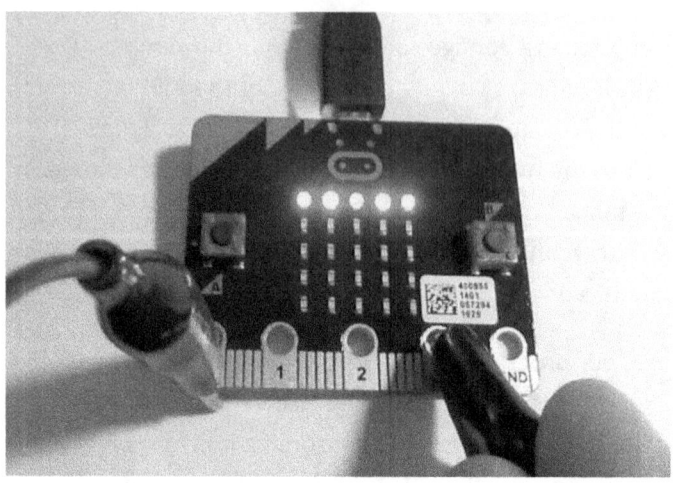

Of course, DC voltage levels can be displayed as well. As expected, you get a horizontal line and can measure the operating voltage of 3.3V, or the voltage 0V on the GND pin.

Our oscilloscope program additionally generates a PWM test output signal with a duty cycle of 50% at the connection point P1. This supplies a rectangular output test signal, which can also be clearly displayed. The standard frequency for this PWM signal is exactly 50 Hz.

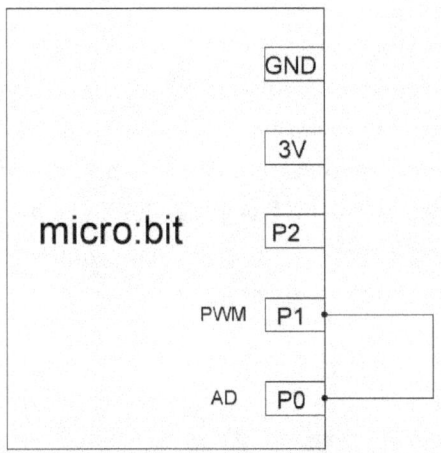

Measurement at the PWM output

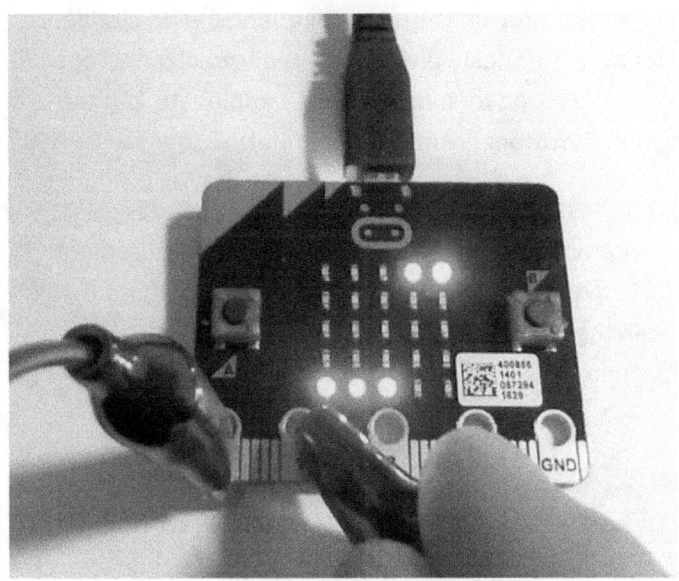

PWM means Pulse Width Modulation and consists of a fixed square wave signal with a switch-on pulse of adjustable length. A connected LED then shows the average level of brightness, which depends on the on/off ratio. A frequency of about 50 Hz is high enough, for our eyes to see no flickering. But if you move the board quickly back and forth, you can see the individual pulses. Our simple oscilloscope shows very well the square-wave signal at the PWM output.

Increasing the sampling rate

Tests using the first oscilloscope software example have shown, that mostly the pause command within the measuring loop determines the timing behavior. As a result, certain timing variations occur in the measurement sequence, which influence timing accuracy. This delay time can be reduced down to one millisecond. The scope then has a faster sample rate, but the measurement is no longer sufficiently uniform.

This looks quite different, when the pause command is completely removed from the loop. The entire measurement loop with its five measuring points then only takes about 0.4 ms. This can then be used to investigate signals within the kilohertz range.

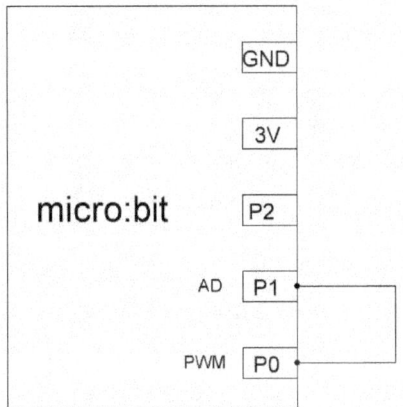

The modified program now uses input 1 and has a signal source of 2 kHz at output 0.

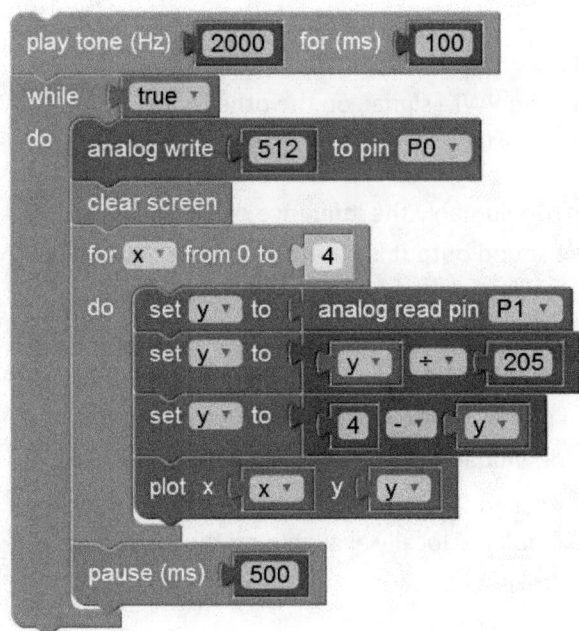

Oscilloscope2.jsz

All sound output commands use output 0. For this reason, the analog input had to be moved to P1. The PWM output is now also at the output 0. More by chance it was discovered that the PWM output takes the same frequency of sound previously generated using *play tone*. This indicates that a timer of the controller is used for this.

The command *set analog period* should actually be responsible for controlling the PWM output. However, at the present state of the software supplied, it was not possible to change the PWM period of 20 ms to any other value. However, it can be replaced by the command *play tone*; this command is very useful, as the actual frequency is sent to the output pin. The generated tone only lasts as long as specified in the command. During this time, the complete program execution is blocked, so unfortunately the generated sound cannot be measured using the oscilloscope. The PWM signal, on the other hand, remains active for an arbitrary length of time and can thus be examined.

A quick warning: presumably the influence of the PWM frequency by the sound output is based on a programming error in the system and could change at any time via a software update. As result, the same program, compiled again using the new update, could behave quite differently. To freeze the current state, all programs are stored as ready-compiled hex files and can be downloaded from the author's website. The additional advantage of such a download: from then on, all the programs in this book are locally available on the PC and no Internet connection is necessary.

```
script Oscilloscope4 (converted)
function main ()
    var y := 0
    var x := 0
    music → play tone(2000, 100)
    while true do
        pins → analog write pin(P0, 512)
        basic → clear screen
        var bound := 4
        x := 0
        while x ≤ bound do
            y := pins → analog read pin(P1)
            y := y / 205
            y := 4 - y
            led → plot(x, y)
            x := x + 1
        end while
        basic → pause(500)
        ♲ micro:bit → pause(20)
    end while
end function
```

Oscilloscope2-converted-2.jsz oder sollte es Oscilloscope4-converted-2.jsz wie es oben im script heisst?

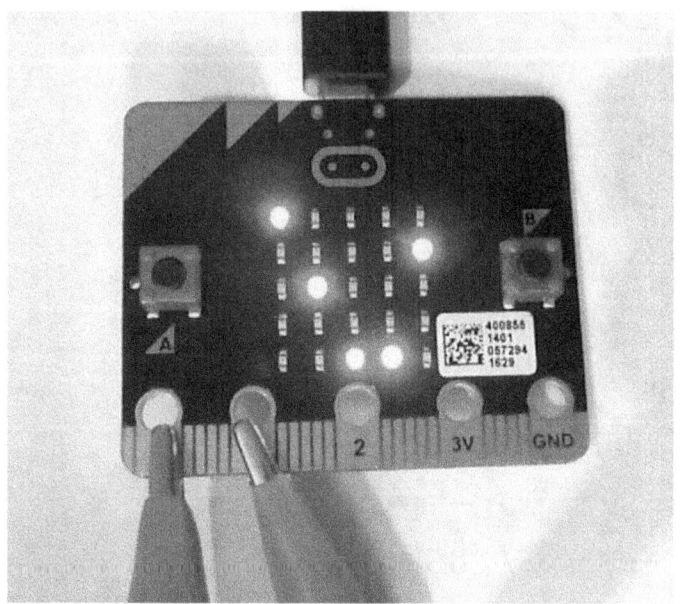

When examining the 2 kHz square-wave signal, slanting edges can be seen, although a faster oscilloscope shows steep flanks. The explanation of this difference: each oscilloscope has a cut-off frequency. Fast signals are generally rounded or flattened at the upper limit. This is another important point which must be considered with each oscilloscope to avoid any mis-interpretation of measurement results.

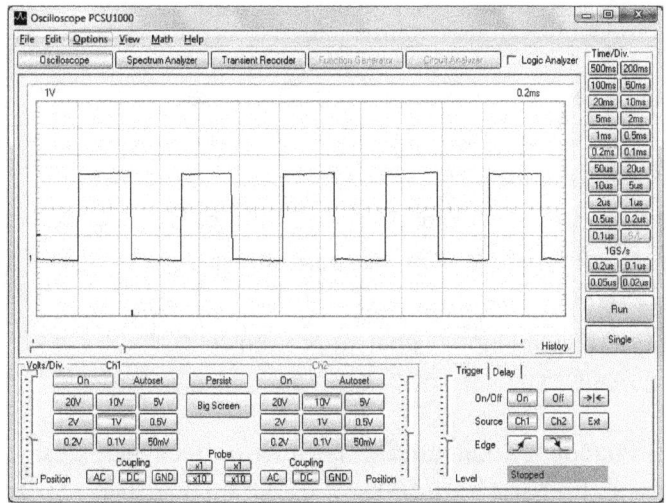

A modern oscilloscope offers a bandwidth of up to 20 MHz or more. In comparison, the AD converter of the Mircro:bit is limited to about 10 kHz. But still, you can measure typical low frequency signals, for example the loudspeaker signal of a music system.

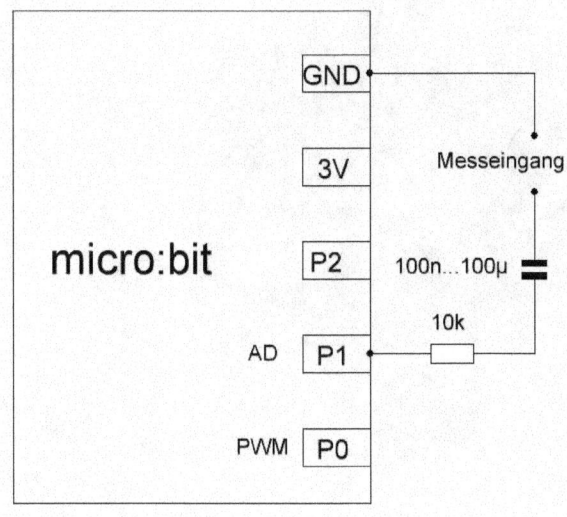

|GERMAN!

But please be careful when connecting to any external signal sources! You must provide the overload protection, because too much input voltage can damage the Micro:bit input or even more. As guidance, a protective series resistor of 10 kΩ and a coupling capacitor of 100 nF or more are sufficient to investigate signals of low frequencies. The capacitor is important to shift the input signal into the correct DC voltage range of 0V and 3.3V.

Such a shift happens automatically, because protective diodes at each input of the microcontroller are built-in; they will cut voltages below 0V or above 3.3V. If an alternating voltage is measured with a peak voltage of - 1 V and + 1 V, the signal will still shift upwards. The following figure shows an oscillogram of a music signal from the headphone output of a radio.

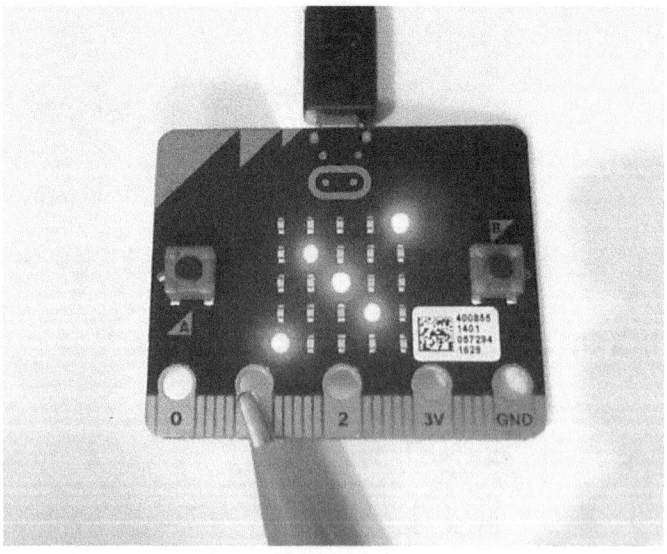

A music signal

Smoothing of a PWM signal

Another important device in any electronics laboratory is an adjustable voltage source or an adjustable power supply. Some devices supply for example voltages from 0 to 30 V and currents up to 5 A. In many cases, much less is needed. A small voltage source of 0 to 3.3 V can be built using the Micro:bit. In this case a PWM signal is smoothed via a low-pass filter, which is built using a resistor and a capacitor.

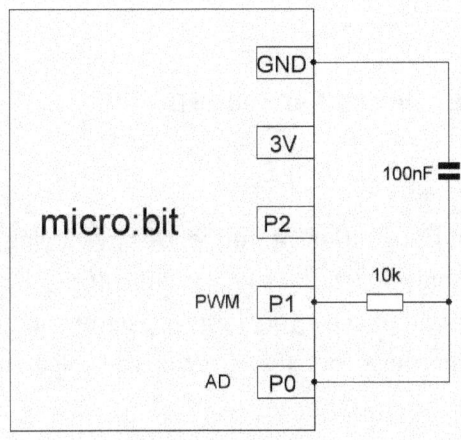

With the help of this low-pass filter of 10 kΩ and 100 nF, the Micro:bit oscilloscope already shows a perfectly smoothed signal as straight line in the center of the screen. At higher resolution, a residual ripple of 0.2 V peak to peak can still be measured. If the resistor is reduced to 1 kΩ, the ripple becomes significantly larger and can also be seen with our simple Micro:bit oscilloscope.

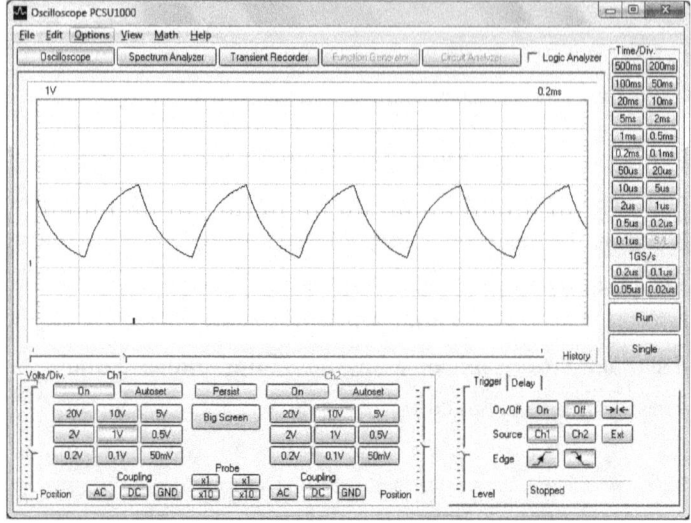

Smoothed ripple at 2 kHz, using 0.1 µF and 1 kΩ

You can also experiment using different frequencies by varying *play-tone* frequency. At only 50 Hz, the signal is difficult to smooth. Then it is necessary to use a very large capacitor such as the 100 µF. To estimate the effect of a filter you can calculate its cut-off frequency:

$f_0 = 1 / (2 * Pi * R * C)$

For 10 kΩ and 100 nF, a cut-off frequency of 159.2 Hz is calculated. On the other hand, an electrolytic capacitor of 100 µF together with a resistance of 1 kΩ result in 1.6 Hz, which is more than 1000 times below the PWM frequency of 2 kHz. Thus, the residual ripple is more than 1000 times less than the amplitude of the unfiltered PWM signal. This is sufficient even for more demanding measuring tasks.

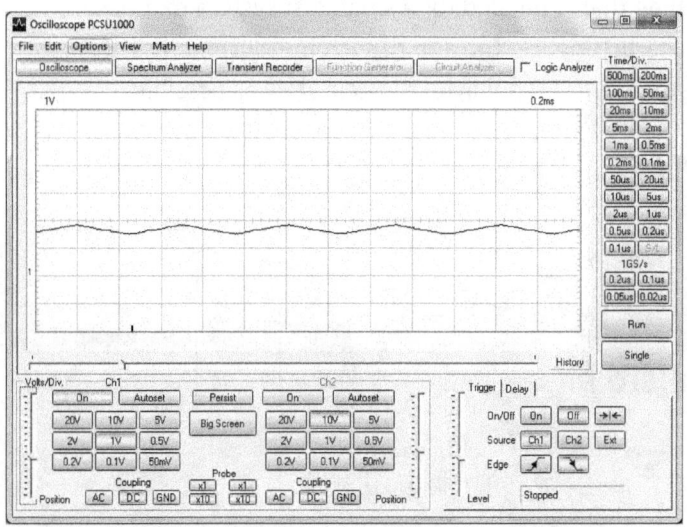

Smoothed ripple at 2 kHz, using 100 nF and 10 kΩ

8 Characteristic curve measurements

Until now, the Micro: bit has already grown into a small measurement laboratory. Even if the oscilloscope only has a rather low resolution, you can still operate the AD converter with high resolution. In addition, there is the PWM output, which also has a high resolution and can supply an adjustable DC voltage via the low-pass filter. This means that we have everything that is needed for a small test system to measure diode characteristics. The aim of such measurements is to understand the LED behavior as accurately as possible.

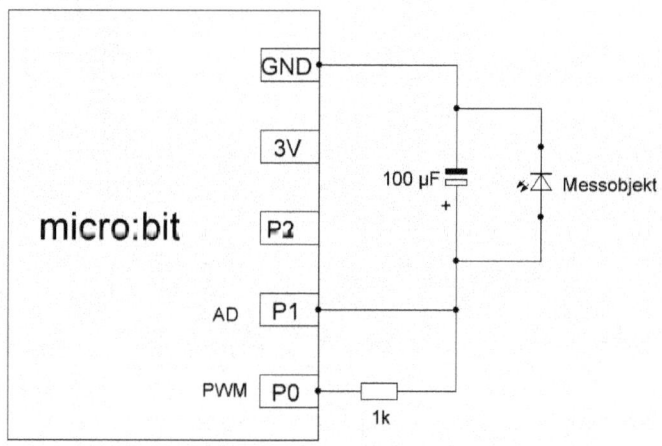

GERMAN Messobjekt

Our measurement setup consists of the PWM output with added low-pass filter, so the voltage is simultaneously connected to the measuring object and the AD input. The output voltage without load can be set to a value of up to 3.3 V. If, however, a current flows through the measuring object, this results in an additional voltage drop at the 1 kΩ resistance of the low-pass filter.

For example, a voltage of 2.3V was set, but only 1.3V is measured at the analog input; then the voltage drop across the 1 kΩ resistor is obviously 1.0V. As we have the 1 V and the 1 kΩ, the current flow of 1 V/1 kΩ = 1.0mA can be calculated.

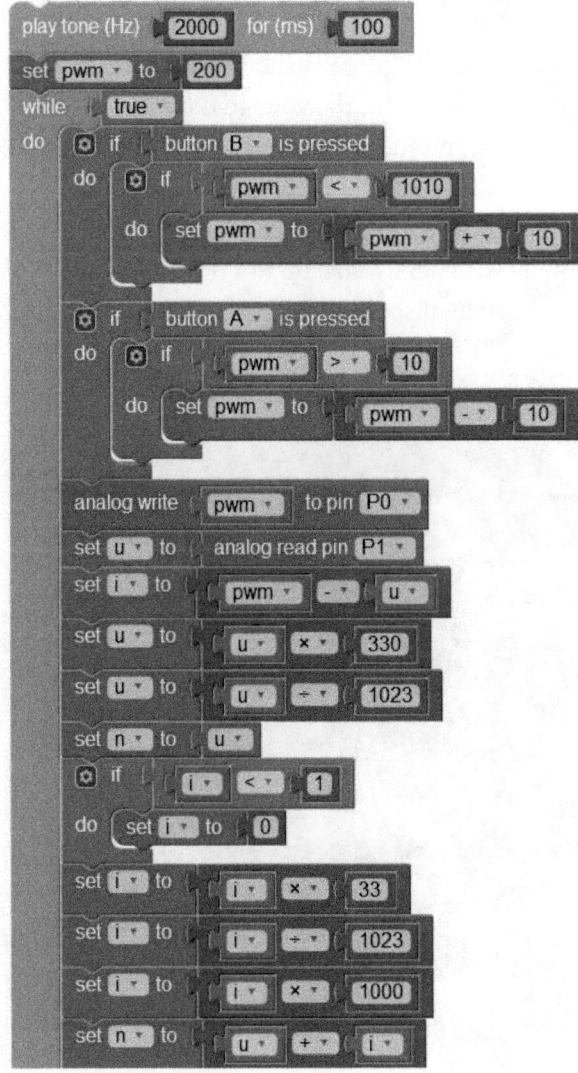

Kennlin1.jsz (shortened) - deutsch aendern? so im Download

This program allows for a very fine adjustment of the output voltage by pressing the switch B (larger) and A (smaller). Taking this voltage setting and the measured voltage at the object under test, we can then calculate voltage U and the current I.

This program uses the same display as shown in chapter 5. The screen is split again. On the right, the voltage is displayed up to 3.30 V using three digits, on the left the current is shown up to 3.3 mA with two digits. For a correct display, the measured values are shifted into the appropriate positions and then added. For example, with 2.55 V and 1.0 mA the number 10255 results, which can then be displayed as usual.

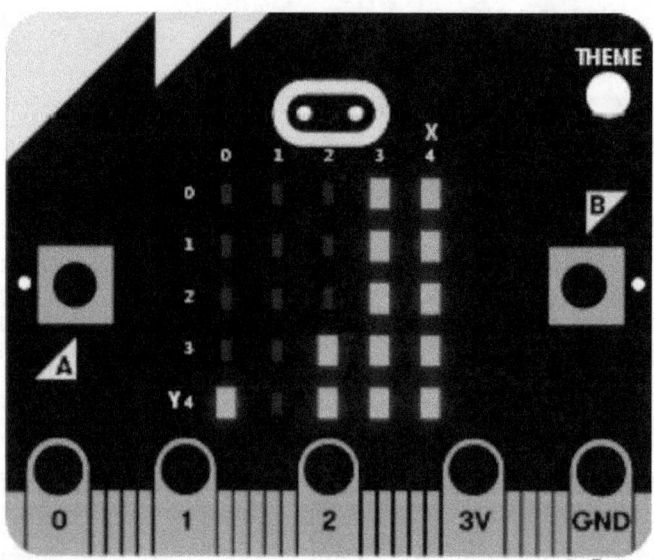

Display of 1,0 mA and 2,55 V

```
pins → analog write pin(P0, pwm)
u := pins → analog read pin(P1)
i := pwm - u
u := u * 330
u := u / 1023
n := u
if i < 1 then
    i := 0
else add code here end if
i := i * 33
i := i / 1023
i := i * 1000
n := u + i
basic → clear screen
```

Kennlinie1-converted.jsz (shortened) ---aendern fuer englisch?

After starting the program, the voltage stabilizes at approximately 0.6 V. If a red LED is connected, it will not light up yet. The voltage must be raised to about 1.6 V using the B switch and then a weak light can be seen. When the output voltage is fully raised to 3.3V, a voltage of 1.78 V and a current of 1.4 mA can be seen on the display.

For the recording of the characteristic curve of this component, you can first increase the voltage until the LED starts to light up. The required current is still below 0.1 mA. From then onwards, the voltage is raised step by step and the voltage and the current are recorded.

Voltage in V	Current in mA
0	0
1,60	0
1,66	0,1
1,68	0,2
1,70	0,3
1,73	0,7
1,76	1,0
1,78	1,4

Using these measured values, the curve of the diode characteristics can then be drawn or generated using a spreadsheet program.

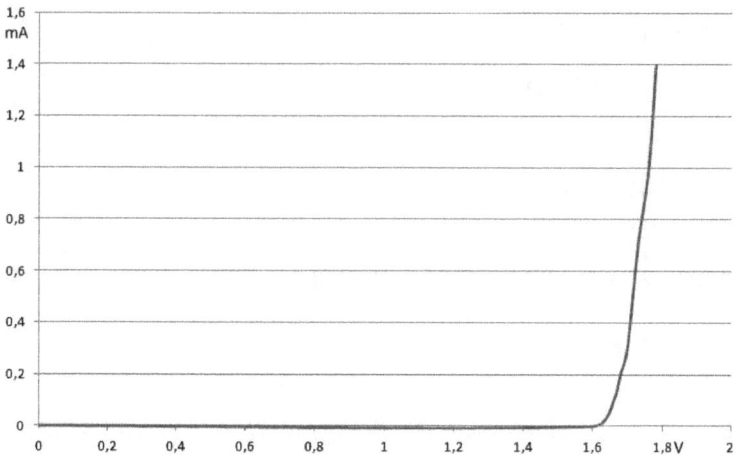

The characteristics of a red LED

The green LED behaves similarly. It is striking to see that green needs an even greater voltage so start lighting up. The characteristics of both are very similar. Up to a certain voltage, no measurable current flows; after this starting point, the current increases steeply. It can be deduced from this, that it is dangerous to connect an excessively high voltage directly. The current would exceed the allowed current limit value of 20 mA.

Voltage in V	Current in mA
0	0
1,75	0
1,84	0,1
1,87	0,2
1,88	0,3
1,90	0,5
1,92	0,9
1,94	1,3

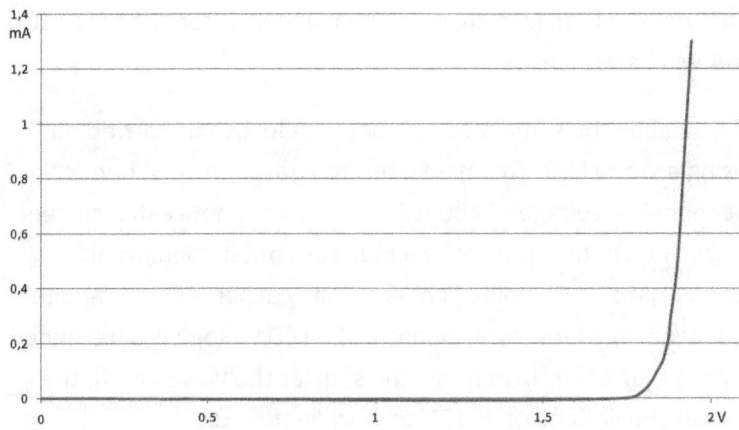

The characteristics of a green LED

But how do these LEDs behave with and without a resistor directly at a port of the Micro:bit? It is well known already, that the internal resistance of the port is about 60 Ω. If the 1 kΩ resistor in the circuit is shortened at the highest voltage output, the current is no longer displayed correctly, and the voltage at the green LED can be measured as 2.2 V. The red LED under the same conditions gave a result of about 1.9 V.

Using the difference to the full operating voltage of 3.3 V, an internal voltage drop in the Micro:bit of 1.1 V (green) and 1.4 V (red) can be calculated. The red LED will then theoretically take 23 mA which can be tolerated, and the green LED is at about 18 mA. However, the actual current through the LED is still significantly lower, which is due to the curved characteristic curve of the FETs (see chapter 6). In fact, a current of about 10 mA will be established.

If available, the same measurement could also be carried out using a white LED. The main point to know before: it needs an even higher voltage of about 2.5 V to start a noticeable current flow. The white LED is really a blue LED, but it contains an additional dye. The blue light is partially absorbed and radiated out again at a larger wavelength. The LED voltage is determined by the blue color. In general, the shorter the wavelength, the higher the voltage of an LED to start lighting up.

Also interesting are measurements on resistors, where a straight-line characteristic is generated.

9 Micro:bit and the Touch Editor

Until now, the Script representation of the Touch Editor was generated by converting the Block based programs, but just to clarify a program. You can also immediately start to program in Script form, leading to some extended possibilities not available before. This relates for example to serial data transmission. Such a communication option with the PC results in extended possibilities for recording and evaluating the measured values.

A new microcontroller has always to prove that the serial interface is working without problems. So, to find out, the Micro:bit web page was searched for examples of serial links. A script in Microsoft Touch Development could be found; it was then extended a bit, and now three analogue values from inputs 0, 1 and 2 can be transferred. In addition, the special software driver had to be installed to establish a virtual serial port for the USB of the PC. And finally, it all works: any terminal using a 115200 bit baud rate can receive the measured values from the Micro:bit.

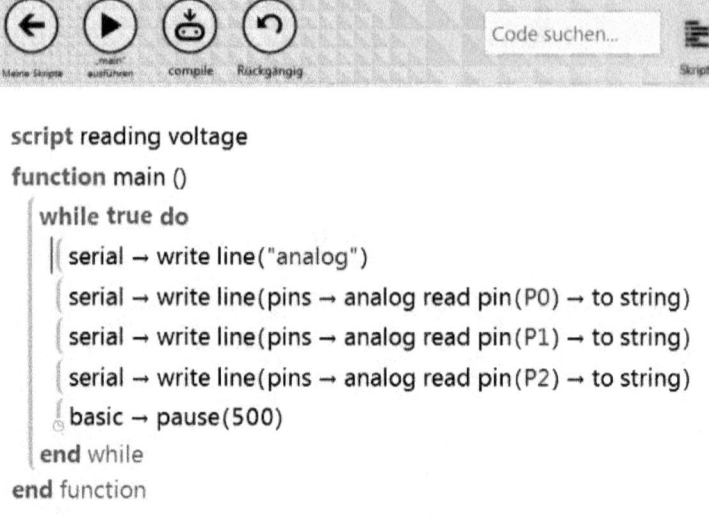

script reading voltage
function main ()
  while true do
    serial → write line("analog")
    serial → write line(pins → analog read pin(P0) → to string)
    serial → write line(pins → analog read pin(P1) → to string)
    serial → write line(pins → analog read pin(P2) → to string)
    basic → pause(500)
  end while
end function

reading-voltage.jsz

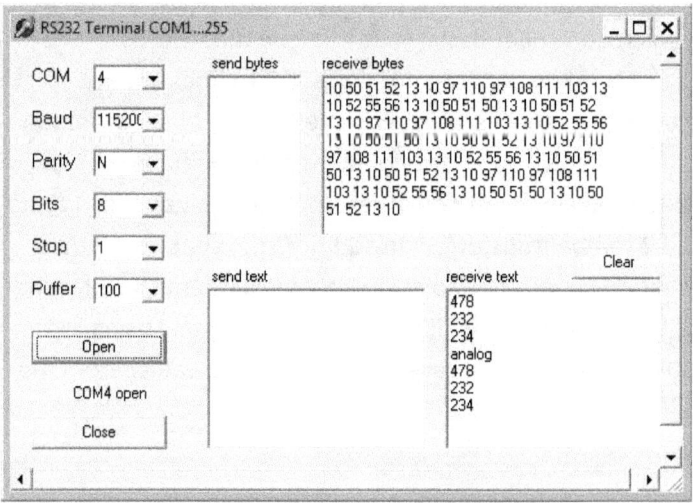

All of the important notes are contained in the documentation for the Serial Library at https://www.microbit.co.uk/td/serial-library. Under Windows you must install the mbed driver on the PC.

For more detailed instructions, see
https://developer.mbed.org/handbook/Windows-serial-configuration. As soon as everything is installed correctly, you will find the mbed serial port in the device manager. After first installation, it had a high COM number, which I then changed to COM4, see screen print.

Screen Print deutsch – ersetzen?--------------------

Using the Microsoft Touch Development Editor, you usually do not type the complete text, but will click on existing keywords. At the end of an existing line, you use Return to create a new empty line. Then, a list of keywords is displayed showing a group containing the desired command.

For example, the *wait command* belongs to the *basic group*. The serial group is at first not visible, it hides behind *More >>*.

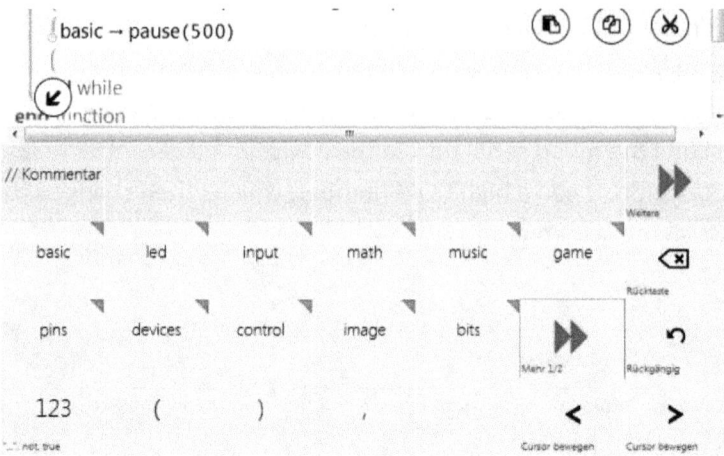

If you click on *serial*, the corresponding functions will appear. In this case, *writeLINE* is used.

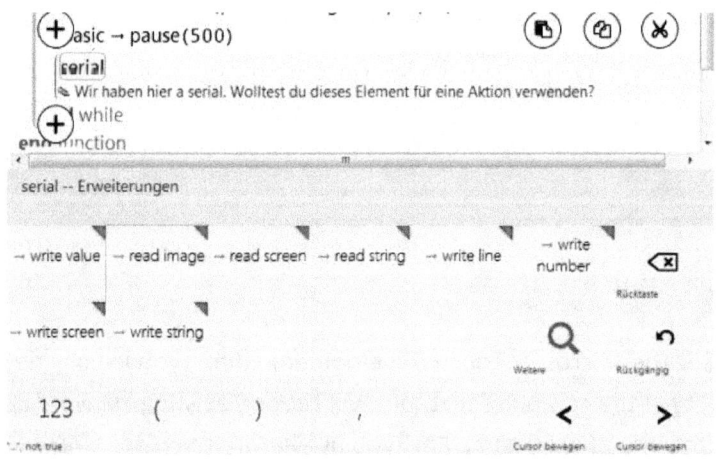

The editor then expects a string of the text that you want to send. Further key words are not suggested or are grayed out. Obviously, a text should now be entered. But actually, we want to send a measured value from the AD converter.

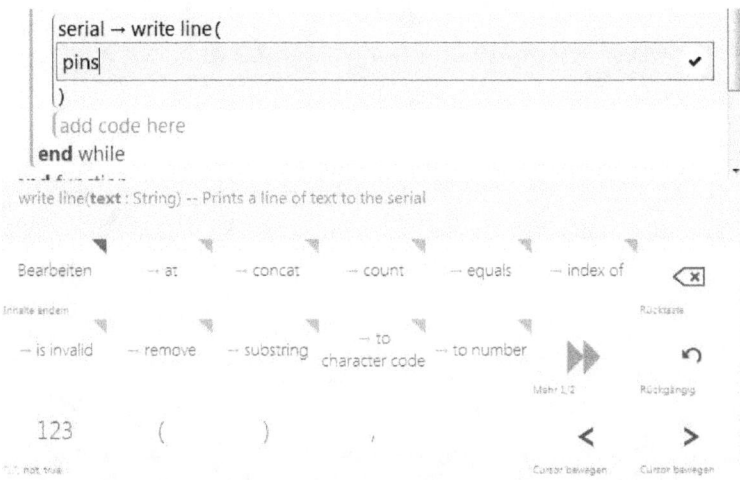

The trial attempt to enter *pins* ends as expected and is interpreted as text to be sent.

Starting from this point, a tedious fight against the editor began, which is probably not intended for such complicated usages. Everything is at the same level as the Blocks, where unfortunately a serial option is not offered at all.

Here, in the Touch Editor, you must remove the quotation mark in front of *pins*, then you can continue; now the functions assigned to the pins appear, including the *analog read pin*. Next you must enter "to" yourself, and as response the type conversion to *string* is suggested.

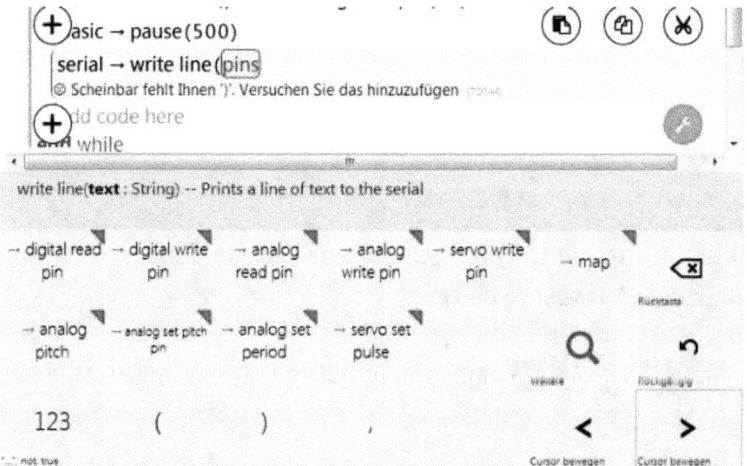

Programming via Microsoft Touch Develop is not easy. But at least the serial interface is now available, and we can transfer any measured values to the PC.

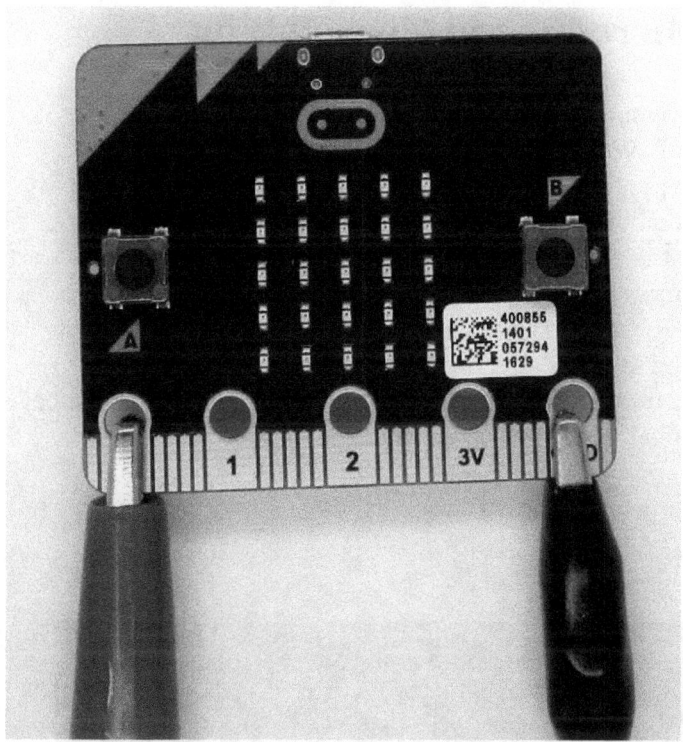

How successful will the Microsoft Touch Development be? Only the future can tell. After all, even Microsoft has now suggested an alternative with PXT, but this is still in beta test phase.

10 Micro:bit and MicroPython

It is very much worthwhile to test MicroPython on the Micro:bit. When you start a new project in MicroPython, a sample program is already suggested that you can test right away. Compilation and downloading works as before, and quickly the example is running. A text scrolls through the LED panel and a heart appears. A very pleasant experience is the fact, that the source text is immediately understandable. You just need to know, that the program structure in Python is created by indenting the lines. Everything that is inside a tabbed block is in this case executed as endless loop.

scroll.jsz

You can find documentation regarding Python at http://microbit-icropython.readthedocs.io/en/latest/index.html. These examples can be directly copied into the editor and tested. Everything runs smoothly and is easy to understand. A sound example has been modified here to understand the sound output code better.

```python
import music

while True:
    music.pitch(20000, 10000)
    for freq in range(1760, 880, -16):
        music.pitch(freq, 6)
```

sound.jsz

Specifically, I wanted to find the highest frequency that can be generated. The maximum frequency is 20 kHz. Using *music-pitch (20000, 10000)*, you produce a sound of 20000 Hz with a duration of 10000 ms. This is then followed by the descending tone as in the original example.

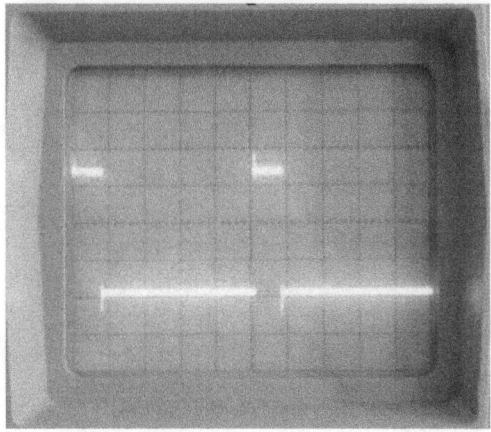

The oscilloscope shows a square-wave signal with additional

short peaks. This means, that there are harmonic waves reaching far into the ultrasound range. If you connect a piezoelectric converter, it could be heard using a bat detector at 40 kHz. Such a sound source is ideal for range and sensitivity measurements. As a result, our Micro:Bit has advanced even further into the noble circle of laboratory measuring instruments.

```python
# Add your Python code here. E.g.
from microbit import *
uart.init()

while True:
    value = pin0.read_analog()
    text = str(value)
    uart.write('U0 = '+ text+ chr(13))
    sleep(200)
```

uart.jsz

And as before, the most important parts are the serial interface and the AD converter. Here, the voltage at pin 0 is measured, converted into a string and then sent serially via the virtual serial interface of the USB. This means, now we have practically all of the tools available to solve even demanding measurement tasks.

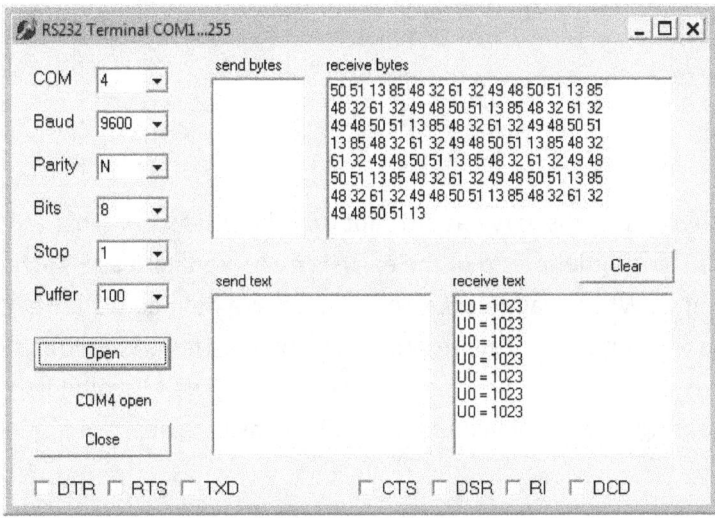

The serial terminal on the PC shows everything as expected, here at a speed of 9600 baud. Here, in this same window, all error messages will be displayed if you made any mistakes.

```python
# Add your Python code here. E.g.
from microbit import *
uart.init()

while True:
    value = compass.get_z()
    text = str(value)
    uart.write('Bz = '+ text+ chr(13))
    sleep(200)
```

compas.jsz

From now on, it is very easy to capture other measurements as well. The magnetic field of the earth is measured here as example. When magnets are approached to a defined distance of 5 cm, then their strength can be compared. This is something that has not been so easy until now in the lab. The Micro:bit is increasingly developing into a versatile measuring device.

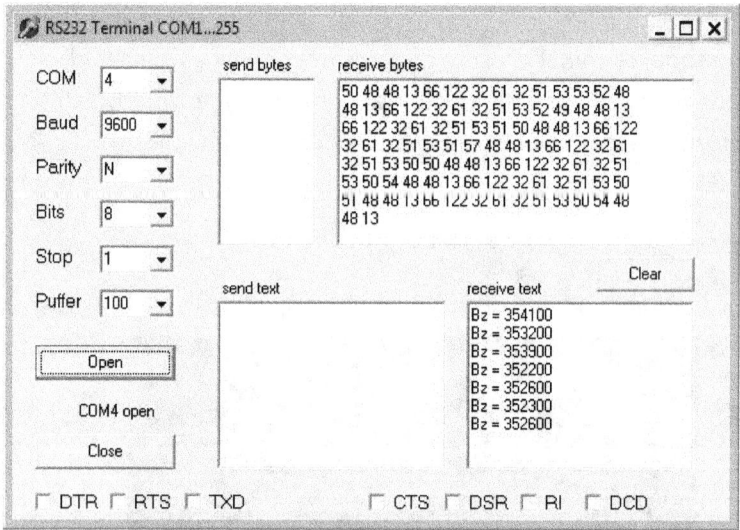

11 Programming Experience Toolkit (PXT)

The Microsoft Programming Experience Toolkit (PXT) is currently only available as the beta version. But a test of this programming environment is worth the time. The Block Editor is here connected to JavaScript. The status of this Beta version can still be seen in many small details, for example, the unusual mixture of German and English terms which hints to the development locations ...

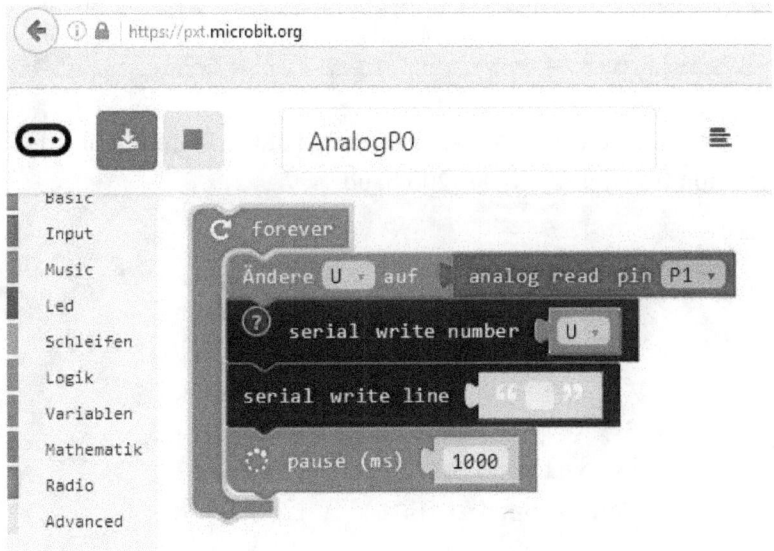

But PXT contains something that was missing in the Blocks: the serial interface via USB. Using *serial write number* and *serial write line,* everything else is very easy. In this case here, an analog value has to be sent to the PC. In this PXT form, the program cannot yet be stored on your own PC. But the hex file can be saved for later use.

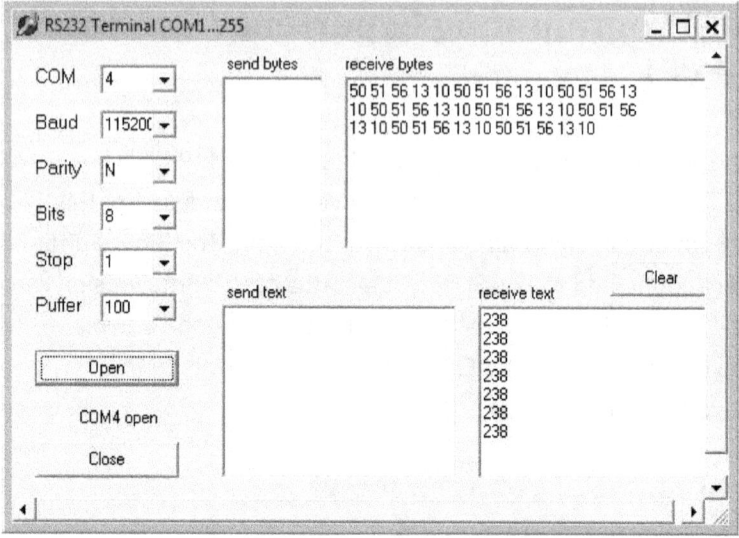

The measured values arrive as expected at the PC terminal. The default transfer rate is 115200 baud.

```
1  let U = 0
2  basic.forever(() => {
3      U = pins.analogReadPin(AnalogPin.P1)
4      // basic.showNumber(U)
5      serial.writeNumber(U)
6      serial.writeLine("")
7      basic.pause(1000)
8  })
```

This program can be converted at any time to JavaScript. In this form, you can copy the text and save it as a text file. A variable is declared at the beginning by the additional row *let U = 0*. The notation for an endless loop *basic.forever (() => {...})* with its many brackets needs a bit of time to get used to. But it does not cause any problems as they are created automatically during the conversion process.

```
let U = 0
basic.forever(() => {
    U = pins.analogReadPin(AnalogPin.P1)
    // basic.showNumber(U)
    serial.writeNumber(U)
    serial.writeLine("")
    basic.pause(1000)
})
```

AnalogP1.txt

As special feature in PXT, you are now allowed to get back into the Block form. And changes to the source code are correctly implemented. In Script form, you can see *basic.showNumber (U)* commented out. This command was originally as Block in our program and had then be commented out in the script. As result it disappeared as Block. Now I can remove the commenting out and maybe change the input channel to P2:

```
let U = 0
basic.forever(() => {
    U = pins.analogReadPin(AnalogPin.P2)
    basic.showNumber(U)
    serial.writeNumber(U)
    serial.writeLine("")
    basic.pause(1000)
})
```

As next step everything converted back, and all of the changes are visible in the Blocks.

At the same time, a Block for declaring the variable U was also added: *Change U to 0*. Writing the code in his way to is unusual, but this may well change again in the next version.

In JavaScript, you can by the way create your own functions. But from then onwards it's over with any backward compatibility; a conversion into Blocks is not possible anymore.

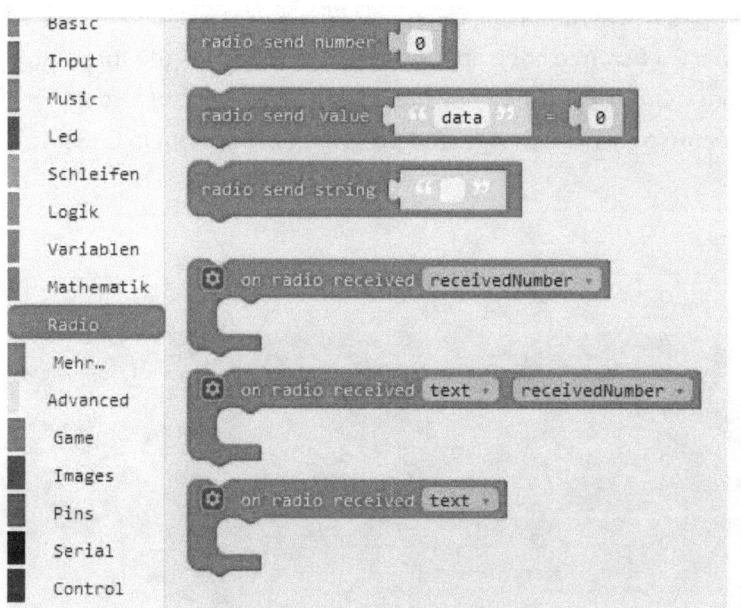

One highlight here is the peer-to-peer radio connection between two or more Micro:bits. The Bluetooth hardware is switched on, but not the Bluetooth protocol. There was something similar already in JavaScript, but now as well in Blocks. An excellent possibility to transmit measured values or control data wirelessly.

12 Programming with Mbed

The entire Micro:bit programming is based on Mbed. When you compile a program using Microsoft Blocks, it is first converted into a C ++ source code and then sent to Mbed via the Internet. From there the finished hex file comes back. You can also go directly to the Mbed page and program your system in C ++.

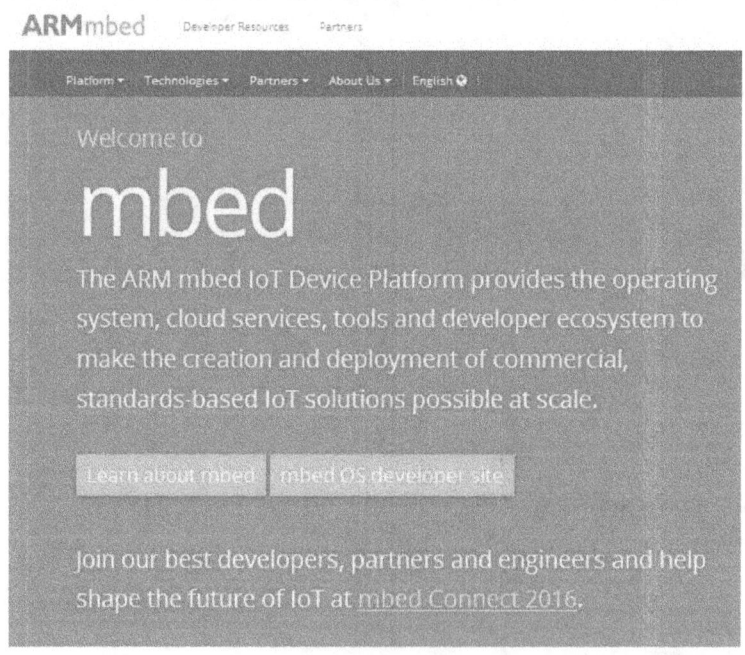

In Mbed, the Micro:bit is just one of many boards. You can therefore treat it like any other 32-bit ARM system.

BBC micro:bit
- Bluetooth Low Energy
- 32-bit Cortex M0, 16MHz
- 16K RAM, 256K Flash

NUCLEO-F746ZG
- Cortex M7 + FPU, 216MHz
- 1-MB Flash, 320-KB SRAM
- USB, Ethernet

In a first attempt, the board was used with mbed.h, deliberately not yet using MicroBit.h. The question was, how fast can you drive a port. During our tests, it became clear, that the controller itself has a port with connections P0: 0 to P0_31, which can also be called p0 ... p31. But this numbering does not correspond with the Micro:bit connections; there are re-sorted. As result I had first to find out, which ports are behind which connections. For the three large ports 0, 1 and 2, these are the ports p3, p2, and p1. Our program example sets all three to outputs and forms a fast loop generating level changes at port 1 (out1). The result is very positive: a port output only needs one microsecond. This then results in a square wave of approximately 500 kHz.

```
/*
Mico:Bit port output 500 kHz
*/
#include "mbed.h"

DigitalOut out2(p1);
DigitalOut out1(p2);
DigitalOut out0(p3);

int main() {
    while(1) {
        out1 = 1;    //1 µs
        out1 = 0;    //1 µs
    }
}
```

Mbed1.cpp.txt

The same example can also be loaded as text from our program archive. It is stored as a pure text file Mbed1.cpp.txt. You can open it using any editor and then copy it into an existing main.cpp.

In a second experiment, the AD converter and the serial interface had to be tested. Again, the connections to the ports have to be re-organized. In Mbed the Mico:bit AD converter supplies a number between 0 and 1.

```
main.cpp  x
1  /*
2  Mico:Bit Analog input
3  */
4  #include "mbed.h"
5
6  AnalogIn voltage0(p3);
7  AnalogIn voltage1(p2);
8  AnalogIn voltage2(p1);
9
10 int main() {
11     while(1) {
12         printf("%f\r\n", voltage1.read());
13         wait(0.5);
14     }
15 }
16
17
```

Mbed2.cpp.txt

The Mbed serial interface software that we had downloaded, installed and used already, works as a virtual COM with 9600 baud. The voltage at pin 1 can be displayed via the terminal. Approximately 0.228 * 3.3 V = 0.75 V is measured at an open input.

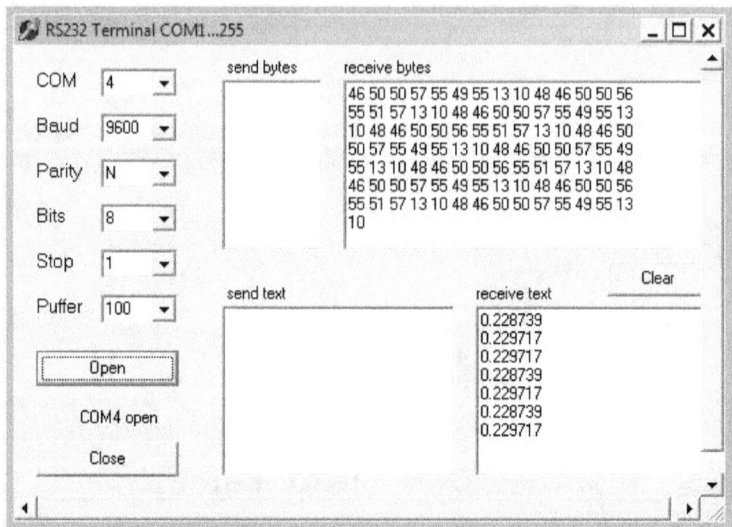

If you test the input using an oscilloscope with a 1:10 probe tip and 10 MΩ, or with the digital voltmeter with 10 MΩ input resistance, a voltage of 1.6 V is displayed. As this is half of the voltage, it can be safely concluded that there are pull-up resistors of 10 MΩ connected to these ports. And in fact, you can also see these resistors at the connections looking at the board. They are probably installed for use of the ports as touch sensors via simple digital queries.

While the average voltage is 1.6 V, the voltage clearly drops to about 0.75 V during a measurement. A capacitor of 100 nF connected to GND helps. Now, almost the full voltage can be measured. In the case of a direct connection to the 3V pin, exactly 1 is displayed.

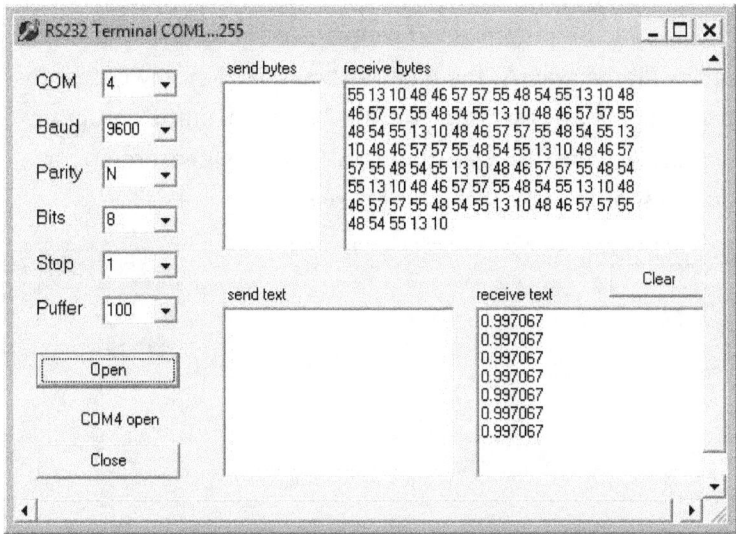

A small program change, and the measured values scaled. The input range is now 0 to 3300 mV.

```cpp
/*
Mico:Bit Analog input
*/
#include "mbed.h"

AnalogIn voltage0(p3);
AnalogIn voltage1(p2);
AnalogIn voltage2(p1);

int main() {
    while(1) {
        printf("%f\r\n", voltage1.read()* 3300);
        wait(0.5);
    }
}
```

Mbed3.cpp.txt

Now a capacitor is connected from analog in to GND and is initially discharged. As the picture shows, it then slowly charges up via the pullup resistor of 10 MΩ. The serial plotter program from the Arduino IDE was used to display the measured data. You can see the typical charging curve.

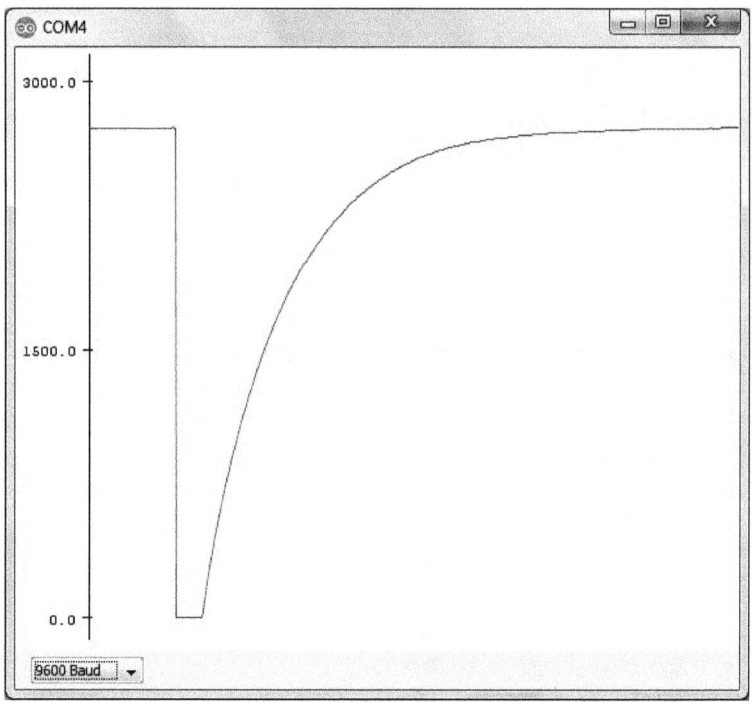

The full circuit diagram of the Micro:bit would help to understand.

If search for a full circuit diagram of the Micro:bit, you find the information that the BBC does not want to publish it yet. This might change after the publication of this book. And until then everyone should know where the different port pins are. To save time for everybody, a program was written which always tests three ports, in this case p4, p5 and p6. The first one generates double pulses, the second triple pulses, and the last four pulses in rapid succession. Using the oscilloscope, the signals can be found quickly, and the relationship port to pin is clear.

```cpp
// Mico:Bit Digital Ports
#include "mbed.h"

DigitalOut o4(p4);
DigitalOut o5(p5);
DigitalOut o6(p6);

unsigned int n;
int main() {
    while(1) {
        for (n=1; n<=2; n++){
            o4 = 1;
            o4 = 0;
        }
        for (n=1; n<=3; n++){
            o5 = 1;
            o5 = 0;
        }
        for (n=1; n<=4; n++){
            o6 = 1;
            o6 = 0;
        }
    }
```

Mbed4.cpp.txt

13 Mbed and MicroBit.h

As soon as you integrate the MicroBit.h file into your own C ++ programs, you are back in the group of developers. Now it does not matter anymore which port of the controller is connected where, you can simply stick to the familiar names. In addition, you have now again the well pre-defined functions, which were already present in the Blocks.

At https://developer.mbed.org/platforms/Microbit/ you can find the official connections and links to further information about the Micro:bit system.

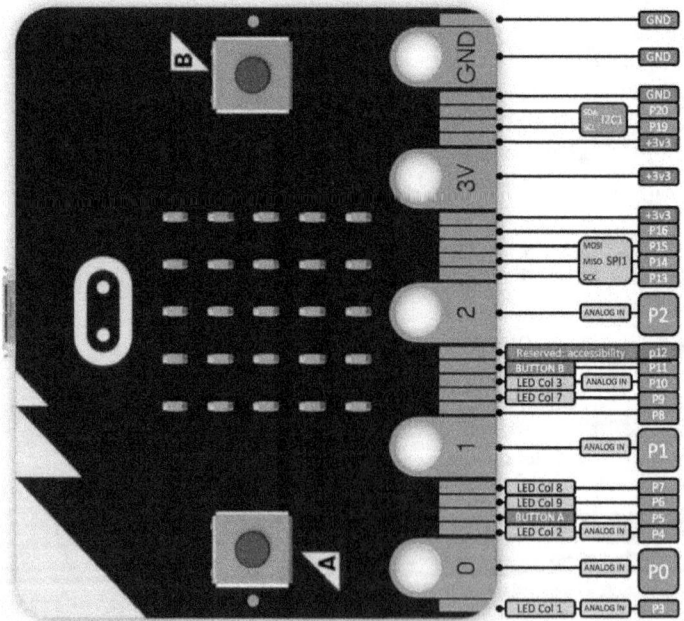

The developers of the Micro:bit at the University of Lancaster have compiled the complete API documentation that you need to use the individual functions. The link is Https://lancaster-university.github.io/microbit-docs/ubit/

With all of this provided, it is not difficult anymore to write the first programs. Here again, one of the first questions was: how does the serial interface work in this environment work? If you have a first running example, everything else then follows very smoothly. The following example measures some values and then transfers them serially to the PC.

```cpp
#include "MicroBit.h"

MicroBit uBit;

int main()
{
    uBit.init();
    MicroBitSerial serial(USBTX, USBRX);
    while (1) {
        uBit.serial.printf("%d\r\n", uBit.systemTime());
        uBit.serial.printf("%d\r\n", uBit.thermometer.getTemperature());
        uBit.serial.printf("%d\r\n", uBit.io.P0.getAnalogValue());
        uBit.sleep(1000);
    }
}
```

Mbed21.cpp.txt

With this Micro.Bit.h code installed, you now have access to parts of the system not available previously. For example, the system time can be read. This is a millisecond based timer restarting with each reset. Also, the internal temperature of the controller is available. Finally, the analog value at terminal P0 is measured. An extra initialization as AD input is not necessary, as the function itself ensures this setting. Different to other functions in Mbed, but the same as in the Blocks, the AD converter returns a 10-bit integer value 0 to 1023.

This example can be loaded as text from our program archive. It is stored as a text file Mbed21.cpp.txt. You can open it with any editor and then copy it into an existing main.cpp. Any terminal can be used to test the program. Here, a transfer rate of 115200 baud must be set.

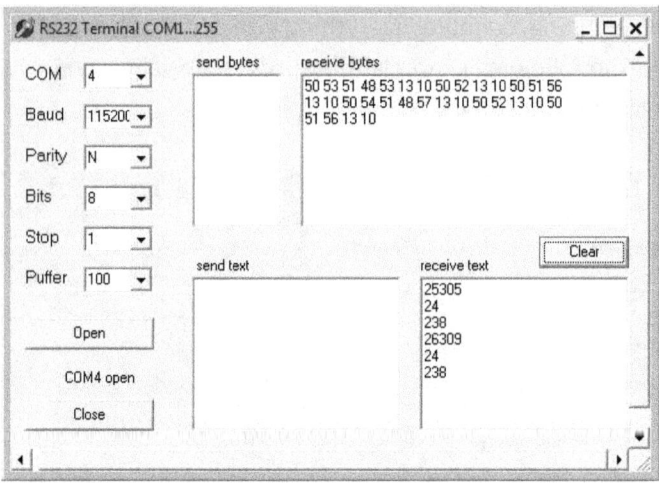

The text window above shows the individual data transfers. This can also be used to measure the delay time and the processing time. There is a delay of 1004 ms between two outputs: 1000 ms of waiting time and 4 ms for the execution of the measurements and the output value transfer.

The temperature measurement shows 24 degrees, and the open-circuit voltage at an open input is shown as 238, the same as already shown in other programming environments. Overall, the behavior when using the MicroBit.h is very similar to that used in Block programming. This is understandable as Block programs are basically translated into Mbed.

The fast oscilloscope

A second program is used to test the output on the LED display and the PWM output. Here, the faster oscilloscope from Chap. 7 is basically re-implemented in C ++.

```
1 #include "MicroBit.h"
2 MicroBit uBit;
3 int main()
4 {
5     int y;
6     uBit.init();
7     uBit.io.P0.setAnalogValue(512);
8     uBit.io.P0.setAnalogPeriodUs(500);
9     uBit.display.enable();
10    MicroBitImage image(5,5);
11    while (1) {
12        for(int x = 0; x < 5; x++){
13            y =   4- (uBit.io.P1.getAnalogValue()/205);
14            image.setPixelValue(x,y,255);
15        }
16        uBit.display.print(image);
17        uBit.sleep(500);
18        image.clear();
19    }
20 }
```

Mbed22.cpp.txt

Again, port 1 is used as analog input and port 0 as PWM output. However, the setting of the period duration now works without any problems and hidden delays. Using a period of 500 µs, the desired output signal of a 2 kHz frequency is generated.

14 Micro:bit short-circuit protection

Some readers have probably already worried about this issue: if the many English students will constantly cause short circuits at the ports, will it thus all go up in smoke. But so far, no black clouds have been seen over the British Isles, and there is a very good reason. I discovered it quite by chance: The developers of the Micro:bit under the leadership of the BBC have installed a short-circuit protection for the ports!

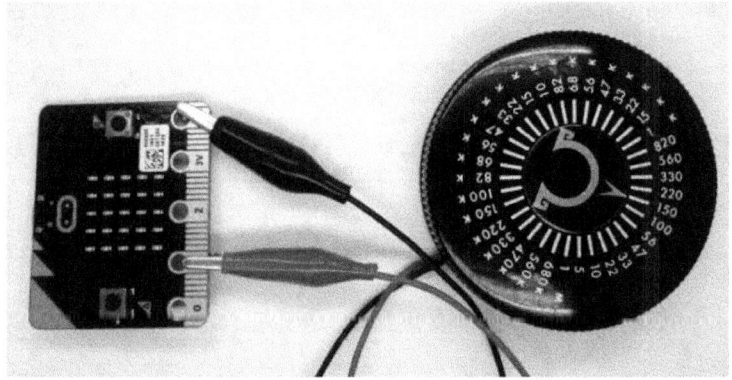

When I wanted to test the load capacity and check the on-resistance of these ports (see chapter 6), I found out about the concept implemented. When I loaded an output with high-impedance like 10 kΩ, everything is still normal. There is an output voltage of about 3.3 V. A heavier load of only 100 ohms was sometimes still ok, but the voltage at the port had then already dropped down to approx. 1.5 V. That means a current of 15 mA, and more is actually not possible! If I increase the load a little bit more, then the port switches off! The first time I thought something must have been destroyed. But after a Reset everything was back to normal.

PortBlink.jsz

For a more detailed investigation, I wrote a short test program; it was designed to generate a square-wave signal at P1. The delay times were only 20 ms. On the oscilloscope, however, I can see that everything runs much slower. The on-time is approx. 12 ms, but the off-time is due to the infinite loop 36 ms long.

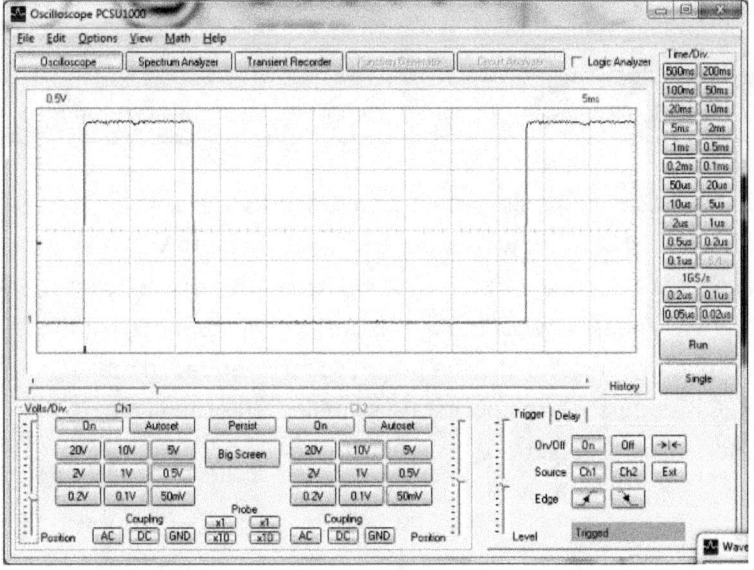

And then it suddenly becomes clearer on the scope: As soon as the port is overloaded, it switched after 6 ms at the latest. Not faster and never much later. The following picture shows the switch-off using 100 Ohm and 15 mA.

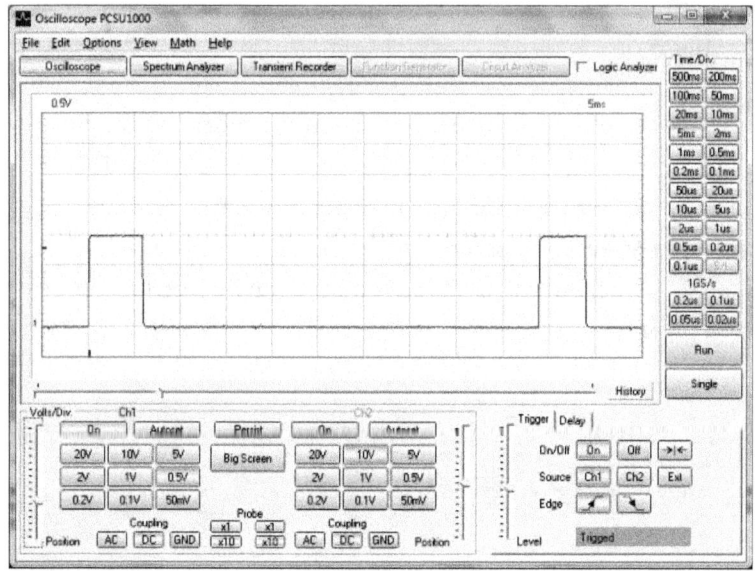

And this is not all: this protection also works in the other direction. When I now connect my load resistor to VCC, there is again the limit at approx. 15 mA. But then the port switches to high! My guess now is: the port is constantly monitored. If I output a high-state, and the port is so far overloaded that 1/2 VCC is undershot, the port state switches. This then works as a short-circuit protection.

Micro:bit – tests, tricks, secrets and code

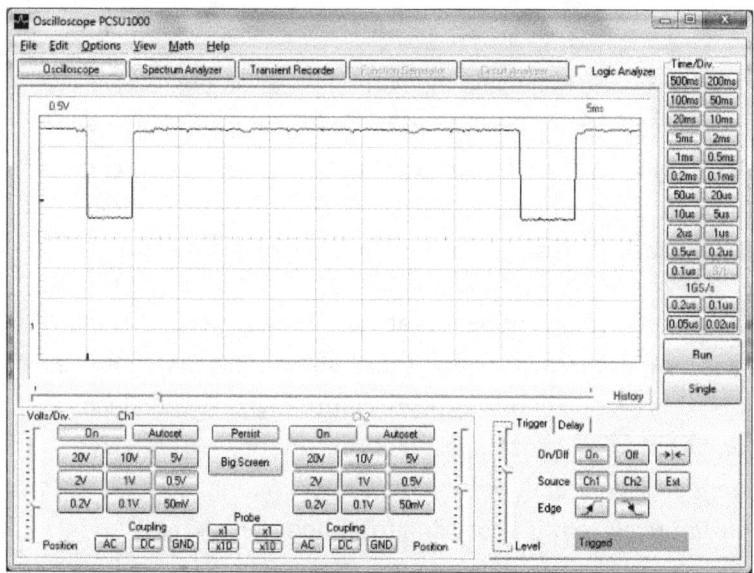

Now, of course, the question arises at which programming level this is implemented? Is this already a property of the controller? Or is it hidden in the Blocks? To find out, a similar test program was written in Python.

```python
# Add your Python code here. E.g.
from microbit import *

while True:
    pin1.write_digital(1)
    sleep(10)
    pin1.write_digital(0)
    sleep(10)
```

port.jsz

As surprise, the test using the Python program does not show any protective reaction. The ports are much faster now, which also has its advantages. But short circuit remains short circuit. The picture shows a load using 47 Ohm. The output voltage has already fallen below 1 V. But there is no trace of a shutdown.

The maximum current here is about 17 mA. This is not a major issue, as per our investigations in chapter 6 this is acceptable. The ports of the controller have a much higher-impedance compared to AVR controllers. For this reason, you can even connect LEDs directly without a current limiting resistor. The maximum current is approximately about 10 mA.

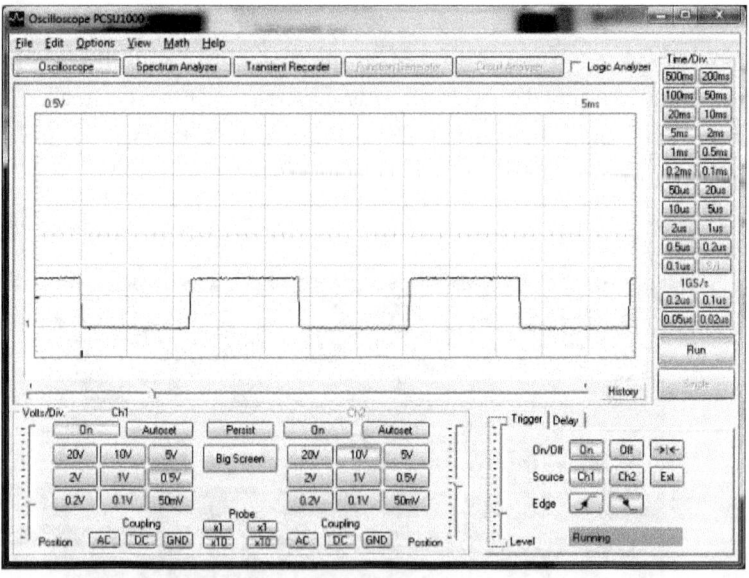

Another test, now using Mbed, shows the same result: No fuse protection function. This experiment was carried out statically with the controller itself measuring the voltage at the port.

main.cpp

```cpp
// Mico:Bit Digital Ports
#include "mbed.h"

DigitalOut o1(p2);
DigitalOut o2(p1);
AnalogIn voltage0(p3);

int main() {
        o1 = 0;
        o2 = 1;
    while(1) {
        printf("%f\r\n", voltage0.read()* 3300);
        wait(0.5);
    }
}
```

Mbed13.cpp.txt

To find out what happens, another test: could it be that by using MicroBit.h port protection is active again? To investigate this question, the same program was tested in this environment.

```cpp
#include "MicroBit.h"
MicroBit uBit;

int main()
{
    uBit.init();
    uBit.io.P1.setDigitalValue(0);
    uBit.io.P2.setDigitalValue(1);
    MicroBitSerial serial(USBTX, USBRX);
    while (1) {
        int u = 3300 * uBit.io.P0.getAnalogValue()/ 1023;
        uBit.serial.printf("%d\r\n", u);
        uBit.sleep(500);
    }
}
```

Mbed23.cpp.txt

But the result this time is again clear: No protection. The port protection is only active when using the Blocks. And this makes

sense: this is the beginner's way of programming. On the other hand, C ++ programmers do not make short circuits ...
And if so, then nothing happens in software ...

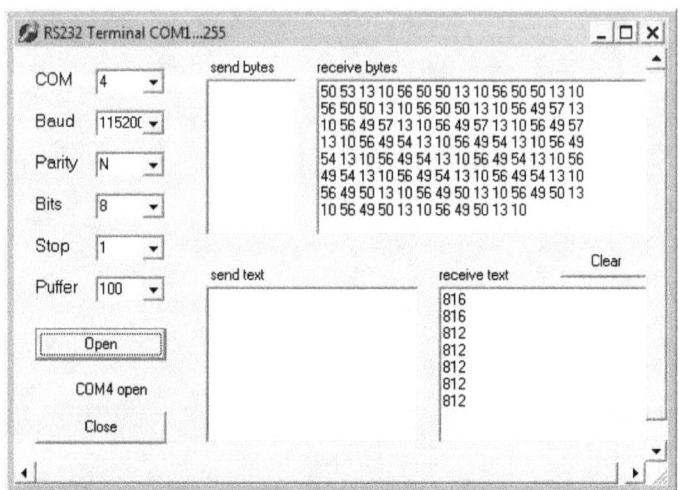

The proof: only 812 mV at 47 Ω

There is no official reference to this protection activity in the Blocks, and as result you can only discover via experiments. Perhaps this was a concession to the Minister of Education, who worried that with the introduction of the 1 000 000 Micro:bits the British schools might burst into flames. On the other hand, they did not want to talk about it too loudly as the pupils concerned might forget about being careful when dealing with electric current. The essential point however is, you can really experiment with little risk and in a very relaxed way.

I hope you enjoyed learning some new aspects about this little beast Micro:bit. News and updates you find on my website.

Burkhard Kainka - December 2016   **version 1.4**

www.ingramcontent.com/pod-product-compliance
Lightning Source LLC
Chambersburg PA
CBHW072212170526
45158CB00002BA/558